T0130741

IOWA'S MINERALS

. . .

A Bur Oak
Original

IOWA'S MINERALS

Their Occurrence,
Origins, Industries,
and Lore

By Paul Garvin

University of Iowa Press ⊔ Iowa City

University of Iowa Press, Iowa City 52242
Copyright © 1998 by the University of Iowa Press
All rights reserved
Printed in the United States of America

Design by Omega Clay

http://www.uiowa.edu/~uipress

Printed on acid-free paper

Library of Congress Cataloging-in-Publication Data
Garvin, Paul
    Iowa's minerals: their occurrence, origins, industries, and
lore / by Paul Garvin.
        p.   cm.—(A Bur oak original)
    Includes bibliographical references and index.
    ISBN 0-87745-626-7, ISBN 0-87745-627-5 (pbk.)
    1. Minerals—Iowa.   I. Title.   II. Series.
QE375.5.I8G37   1998
553'.09777—dc21   97-52230

98 99 00 01 02 C 5 4 3 2 1

98 99 00 01 02 P 5 4 3 2 1

This book is dedicated to the generations of mineralogy students at Cornell College, whose enthusiasm for learning has been a driving force in my study of Iowa's minerals.

Contents

Acknowledgments

This book would not have been possible without the generous assistance of many individuals and organizations. Jean Prior gave much encouragement from the very beginning of this undertaking. Raymond Anderson, Raymond Rogers, and my wife, Ellen Garvin, read the manuscript and made numerous suggestions for improving it. Raymond Anderson, Robert Cody, Mary Howes, Robert McKay, and Jean Prior shared valuable information concerning Iowa's mineral industries and mineral lore. Ed Clopton, Dennis Kossow, David Malm, George McCormick, and Karl Seifert provided information about specific minerals and mineral occurrences. Tony Plaut and Dale Beeks supplied photographic equipment and expertise in the production of the mineral photographs. Mary Bennett and other staff members at the State Historical Society of Iowa, Pat Lohmann of the Geological Survey Bureau, and Julia Goldin of the Department of Geology at the University of Iowa helped to identify written and photographic material for inclusion in the book. Paul Van Dorpe supplied photographs related to coal mining.

John Cordell and the staff at the Office of the State Archaeologist of Iowa provided Native American artifacts for photographing. Several producers of mineral and rock products provided written and photographic materials, as well as permission to enter quarries and mines to gather information on minerals. These producers include: Basic Materials Corp., Kaser Corp., Lafarge Corp., Linwood Mining and Minerals, Inc., Martin-Marietta Aggregates, River Products Co., United States Gypsum Corp., and Wendling Quarries, Inc. Iowa Limestone Producers provided additional information on the limestone industry.

Gratefully, I acknowledge the assistance of all those listed. Funding for this book was provided, in part, by grants from Cornell College, the Iowa Science Foundation (Grant #ISF-97-28), and Linwood Mining and Minerals, Inc.

Preface

Almost everyone has an interest of some kind in minerals. We depend upon minerals because they are the chief components of a wide variety of useful commodities, such as plaster, ceramic, concrete, building stone, and road-surfacing materials. Mineral industries are a vital part of the economy of every state in the nation and of virtually every country in the world. We spend considerable sums of money in order to adorn ourselves with mineral gems. Some people are attracted to the natural beauty of minerals and simply collect them. That minerals have status in Iowa is evidenced by the fact we have our own state rock—the quartz geode. However, the value of minerals is more than monetary and aesthetic. Minerals are the fundamental building blocks of which Iowa and the rest of the Earth are made; therefore, understanding their natures and origins is an important prerequisite to interpreting Iowa's geologic history. Indeed, minerals are essential elements of the language in which that book of history is written.

My interest in Iowa's minerals began in 1972, when I had the good fortune to attend a Tri-State Geological Field Conference. The conference, sponsored by the Department of Earth Sciences at the University of Northern Iowa, featured a trip to Pint's Quarry, a short distance east of Waterloo and at the time one of the premier localities in the state for collecting minerals. As I examined cavities in the dark brown limestone and the crystals of calcite, pyrite, fluorite, and barite that lined them, I began to envision a real opportunity to engage my mineralogy students at Cornell College in a firsthand study of minerals. The first student research project was a comparative study of the mineral deposits at Pint's and Conklin quarries. In the years that followed, other deposits in eastern Iowa were investigated, with the result that Cornell students and I have collected and studied many of the mineral deposits in Iowa. The purpose of the research has always been to increase our understanding of the geologic history of Iowa's minerals.

Nearly twenty-five years ago Paul Horick, then of the Iowa Geological Survey, wrote *The Minerals of Iowa* for the survey's Educational Series. Designed primarily for use by elementary and secondary school students and teachers and by mineral collectors, the book has served well in that capacity. Since 1974 several new mineral deposits have been discovered and much more has been learned about the origins of minerals in Iowa, but the information is scattered in professional journals, field trip guidebooks, and state government publications. I believe it is time for a new book.

Although the variety of minerals native to Iowa is rather limited, mineral specimens of exceptional quality have been and continue to be found by researchers and collectors. The vast majority of the museum-quality specimens come from carbonate-hosted epigenetic deposits in the eastern part of the state. Iowa's mineral elite reside in museum and private collections in Iowa and elsewhere. They have been featured in the *Mineralogical Record*, a magazine for mineral collectors and scientists that features photographs of high-quality specimens (Anderson and Stinchfield 1989; Garvin and Crawford 1992). Appendix C lists museums in Iowa where mineral specimens from the state are displayed. In selecting minerals for the color plates in this book, I have drawn primarily from the Anderson Museum at Cornell College and from my own private collection. I have attempted to feature minerals from a variety of occurrences in the state, though the Linwood Mine has by far the largest representation. Criteria for inclusion are mineral variety, freedom from damage, and aesthetic characteristics, such as size and patterns of shape and color. I chose not to use thumbnail specimens or micromounts. The specimens featured in the color plates are merely a sampling of the many beautiful minerals that have been collected from quarries, mines, and outcrops in Iowa.

This book is written for students, educators, and collectors of minerals. It also explores the fascinating history of Iowa's mineral industries and its mineral lore. Its purpose is to broaden our awareness of, and appreciation for, Iowa's mineral endowment and to deepen our understanding of how that endowment came to be.

IOWA'S MINERALS

1. Origins of Iowa's Minerals

All of the solid material of the Earth, including rocks and unconsolidated sediment but excluding volcanic glasses, contains minerals. The moment you leave your house you encounter minerals. They are present in the concrete of your sidewalks and in the rock salt used to remove ice from them. In fact, ice itself is a mineral. Minerals are found in paved roads and in the gravel on unpaved roads. They are present in the stone that forms the foundations, lintels, and in some cases the walls of many old homes and barns in Iowa. They are the stuff of which tombstones and other monuments are made. They are found in floors, steps, walls, columns, and counters of libraries, office buildings, and banks. Take, as one example, the Norton Geology Center at Cornell College in Mount Vernon, Iowa. The exterior steps and belt courses around the building are made of tan limestone. The interior stairs are mottled white and gray marble. A lecture room countertop is dark green soapstone. The chalkboards and one set of stairs are black slate. A sundial base, a short distance west of the building, is sculptured from pink granite. Add to these the drywall, plaster, and concrete, all prepared from minerals, and it is easy to see that minerals are all around us.

What Is a Mineral?

In scientific terms, a mineral is a naturally occurring, chemically homogeneous crystalline solid. "Naturally occurring" means that a mineral is not the product of an artificial process. It is possible, however, to synthesize most minerals in laboratories. For example, the gem minerals diamond, ruby, and emerald can be created in laboratories. When the term "synthetic" or "created" precedes the mineral name, it means that the substance was produced artificially. Many chemical substances have been synthesized, for which there are no known natural counterparts. Such substances are not given mineral names. "Chemically homogeneous" means that the chemical composition is defined sufficiently to be expressed by a chemical formula. For instance, the chemical composition of the minerals quartz and calcite are represented by the formulas SiO_2 and $CaCO_3$, respectively. The chemical formulas of some minerals are long and appear complex, reflecting, in part, that their compositions are variable. "Crystalline" means that the chemical constituents are symmetrically distributed in a three-dimensional structure (called a crystal lattice). Minerals are, most often, products of inorganic processes; however, some minerals result from the action of living organisms. Calcite, for example, is secreted by invertebrate animals, like clams and mollusks, in the manufacture of external shells. The bones and teeth of vertebrates are made of the mineral apatite. The activity of certain bacteria can cause precipitation of minerals like pyrite.

In a court of law, the word "mineral" is defined differently than it is in a scientific laboratory. In a legal sense, a mineral is a naturally occurring substance that has sufficient economic value in its situs (natural setting) to be exploited profitably. Thus, petroleum, natural gas, and water may be considered mineral resources, despite the fact that they are not solids, and granite may be a mineral, even though it is not chemically homogeneous.

The Difference between Rocks and Minerals

The terms "mineral" and "rock" are often used interchangeably in everyday conversation. Technically, they are not the same. All rocks, except volcanic glasses, are aggregates of grains or crystals of one or more chemically homogeneous minerals. The aggregation is a chemically inhomogeneous rock. When considering the origins of rocks and minerals, the distinction between the two is less important. Any discussion of the origin of rocks requires an understanding of the minerals they contain; any discussion of the origin of a given mineral requires a consideration of the other minerals with which it is in contact.

The minerals that occur most abundantly at the surface in Iowa are those that make up the sandstone (quartz), shale (quartz and clay minerals), limestone (calcite), and dolostone (dolomite) bedrock and the soils and alluvium (primarily quartz and clay minerals) that blanket them. The generally fine-grained minerals in these rocks and regolith do not attract the typical mineral collector because they lack color and luster (light reflectivity), properties that endow a mineral with natural beauty. For most minerals, color and luster intensify as the size of a mineral crystal increases; therefore, mineral collectors seek occurrences of large crystals (greater than a few millimeters in length). Most often, large crystals are found in natural openings in otherwise solid rock. In Iowa, such openings commonly occur in carbonate rocks, because these rocks are susceptible to dissolution by underground water. Conditions that favor the production of large, well-formed crystals include freedom from interference during growth and a slow growth rate.

Mineral-Forming Environments

Observing the glasslike transparency of a Brazilian quartz crystal, the blood red color of a Burmese ruby, or the mysterious fire of an Australian opal prompts the question: how did these minerals and their attractive qualities come to be? Many people, when observing the lustrous, well-developed faces of a natural crystal, are amazed to learn that it was not machined by a lapidarist. The crystal faces and their symmetrical arrangement are, in actuality, an outward expression of the internal arrangement of ions, atoms, and molecules that are held together by forces of attraction called chemical bonds. Minerals grow; in fact, for centuries it was a commonly held belief that they, like animals and plants, were alive. Some natural philosophers believed that they resulted from some kind of sexual union deep in the bowels of the Earth. We now know that minerals grow not by internally driven reproduction but by the addition of material from an external medium. Growth occurs outward from the core of the mineral, which is called the nucleus. Nucleation and growth may occur in any of the following media: a molten liquid, usually of silicate composition; an aqueous (water-rich) fluid; a vapor; or an enclosing solid rock. The fluid may be at surface temperature or heated (up to 600°C or more). Identifying the causes of mineral growth requires a thorough study of mineral structure and chemistry. Interpreting mineral structure leads to an understanding of the physical conditions of growth. For instance, carbon exists in nature in two common structural forms: graphite and diamond. Pressure/temperature studies of these minerals reveal that graphite is the stable form of carbon at low pressure, whereas diamond requires high pressure for stability. Pressure in the Earth increases with depth below its surface, from which we deduce that diamonds form at a great depth. Chemical analyses of minerals lead us to an understanding of the chemical environment of mineral growth. Knowledge of the content of certain trace elements or isotopes in a mineral helps us to identify plausible sources of mineral-forming chemicals and to interpret the chemical changes that might have occurred during the process of growth. For example, an analysis of the isotopic composition of sulfur in a mineral sulfide (like pyrite) or a sulfate (like barite) can help us determine whether the sulfur might have come from sea water or from a cooling magma.

Growth layers (zones) within a crystal provide evidence of mineral growth as a function of time (fig. 1.1). Chemical changes that occur during growth can be determined by performing detailed chemical analyses of each zone. Minerals rarely form as isolated individuals; rather, they usually form as part of an assemblage consisting of several minerals that are

1.1. Growth zones in calcite crystal. Robins Quarry, Linn County. Specimen is 13 centimeters long. *Anderson Museum, Cornell College.*

products of the same mineral-forming event. Analyses of the component minerals and the determination of the chronological order of their appearance aid us in tracing the history of the assemblage. Determining the origin of minerals leads to an understanding of the origin and history of the Earth. Indeed, minerals are an indispensable part of the language of rocks from which the history of the Earth can be read—if the reader can interpret the language.

In terms of their modes of origin, all minerals belong to three broad groups: igneous, sedimentary, or metamorphic. Igneous minerals are products of the crystallization of magma, which originates at least several kilometers beneath the surface of the Earth, where heat is sufficient to melt solid rock. The magma may crystallize at or near its place of origin, it may ascend along fractures or other avenues of weakness in the overlying rock to shallower levels in the Earth where it may pool and solidify, or it may issue forth as lava and solidify at the surface. The chemical composition and amount of rock melted, the changes that occur in the magma during its ascent, and the place and conditions of solidification combine to determine the identities of the resulting minerals. Because the rocks that are melted are almost invariably rich in silicon and oxygen (the two most abundant chemical elements in the Earth's crust and mantle), igneous minerals are predominantly silicates. The most common silicate minerals also contain varying amounts of aluminum, iron, magnesium, calcium, sodium, and potassium—elements that are also relatively abundant in the

Earth's crust and mantle. Most igneous minerals belong to one of the following silicate mineral groups: olivine, pyroxene, amphibole, mica, potassium feldspar, plagioclase, or silica (quartz). For example, granite contains chiefly potassium feldspar, plagioclase, and quartz, while basalt is primarily plagioclase and pyroxene. Under special conditions of crystallization, some magmas may also produce rather unusual and valuable gem minerals like topaz, tourmaline, amethyst, and beryl (aquamarine and emerald). The identities of the minerals present in the igneous rock, their chemical compositions, and the sizes, shapes, and arrangements of their crystals help geologists determine the source of the magma and the conditions under which it solidified (fig. 1.2).

Sedimentary minerals result from processes that operate at or near the surface of the Earth. Some minerals are, in a sense, not sedimentary because they are remnants of preexisting rocks that have survived physical and chemical disintegration caused by weathering. Examples are quartz, magnetite, garnet, and zircon. Others, like the clay minerals hematite and limonite, are formed through physical and chemical decomposition of preexisting minerals. Particles of loose sediment are transported by an agent, like running water, wind, or glacial ice, to a place of deposition, such as a streambed, a desert floor, or the margin of an ice sheet. The accumulated sediment eventually becomes lithified (turned to stone) through compaction, cementation, or recrystallization. As an example, quartz sandstone is a sedimentary rock composed of quartz grains that were released by weathering from a source rock (e.g., granite, quartzite, or a preexisting sandstone) and subsequently transported, deposited, and cemented together. Sedimentary minerals may also form through biological or nonbiological precipitation of chemicals that were dissolved in water.

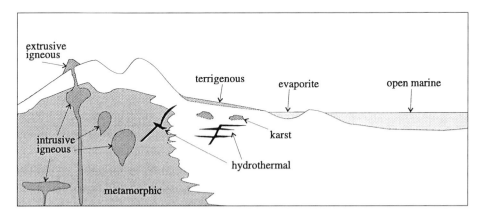

1.2. Generalized cross section illustrating mineral-forming environments: igneous (intrusive and extrusive), sedimentary (terrigenous, evaporite, open marine, and karst), metamorphic, and hydrothermal.

Dissolved mineral matter gives natural water its taste. These chemicals are derived primarily through the chemical weathering of source rocks. Underground and surface waters carry dissolved material to areas suitable for precipitation. Precipitation may occur in terrestrial environments—for example, in air-filled caves (e.g., stalactites), through chemical replacement of plant material (e.g., petrified wood), or in small water-filled cavities in limestone (e.g., geodes). Minerals may also precipitate in marine environments—for example, in shallow marine basins that harbor abundant life (e.g., calcite and aragonite) or in isolated coastal lagoons (e.g., evaporite salts like gypsum and halite). Some precipitation is mediated by living organisms (e.g., the secretion of calcite or aragonite to make skeletal materials in brachiopods and mollusks and bacteria-assisted precipitation of pyrite in coal and black shale). After sediments are deposited, they may experience physical and chemical changes during burial, giving rise to another class of sedimentary minerals referred to as diagenetic. Examples of sedimentary diagenesis include replacement of calcite in limestone by dolomite, replacement of limestone by chert, and formation of gypsum crystals in shale through decomposition of pyrite (fig. 1.2).

Metamorphic minerals form deep beneath the surface of the Earth when heat, pressure, and chemically active fluids combine to convert unstable minerals in a preexisting rock (protolith) to new ones. For example, simple recrystallization of the mineral calcite in limestone produces marble, thermal decomposition of iron-rich shale may generate micas and garnet (schist), and heat and water expelled from a cooling magma may convert limestone into a rock that is rich in calcium silicate minerals (skarn). The temperature boundary between burial diagenesis and low-intensity metamorphism is not sharply defined, but it is generally considered to be in the range of 200 to 300°C. Which metamorphic minerals form in a given deposit depends primarily upon the composition of the protolith, the temperature and pressure of metamorphism, and the activity of dissolved chemicals in water interacting with the rock. Since most protoliths consist mainly of silicate minerals, metamorphic minerals are mainly silicates. Most major silicate minerals in igneous rocks are also found in metamorphic rocks. Other common metamorphic minerals include chlorite, garnet, staurolite, kyanite, sillimanite, talc, diopside, and actinolite (fig. 1.2).

Minerals that form from the action of hot subsurface water are called hydrothermal minerals. The water becomes hot (at least 50°C) by flowing through rock openings in close proximity to bodies of cooling magma or by descending to depths in the Earth where all rocks are naturally hotter. Heated waters dissolve chemicals from the rocks through which they pass. Precipitation of minerals occurs as the waters cool and become oversaturated with dissolved chemicals or as the waters react with the surrounding rocks. Common hydrothermal minerals include many of the mineral sulfides, such as pyrite, galena, sphalerite, and chalcopyrite. Non-sulfide minerals, such as quartz, calcite, and barite, may also form hydrothermally (fig. 1.2).

One other category of minerals deserves mention: minerals that are products of meteorite impact. These minerals are found in meteorites and tektites. Meteorites are extraterrestrial rocks that have reached the Earth, the vast majority of which are believed to come from the asteroid belt, a region of planetary debris that orbits the Sun between Mars and Jupiter. Detailed chemical analyses provide evidence that certain rare meteorites originated on Mars and the Moon. Compositionally, meteorites are of three types: irons, stones, and stony irons. The irons consist chiefly of iron-nickel–rich metallic minerals, including kamacite (iron) and taenite (nickel-iron). They are heavy and typically are coated with a layer of rust-colored iron oxide. The stones are primarily mafic silicate minerals, like pyroxene and olivine, and they resemble closely some terrestrial igneous rocks. Stony irons contain both metallic and silicate minerals. Meteorites may be found through observation of their impact (falls) or by discovery after impact (finds). Because of their similarity to terrestrial rocks, stones are generally identified only as falls, whereas irons or stony irons can quite easily be identified as finds because they are unlike Earth materials. Well-known meteorite fields include those at Meteor Crater near Tucson, Arizona; at Plainview and Odessa, Texas; and at Henbury, Australia.

Tektites are small, typically spheroidal glassy objects with an extra-terrestrial connection. Their surfaces are generally pitted. They are chemically differentiated from glassy terrestrial material (obsidian) by their very low water content. Large tektites are typically spheroidal, button-, tear-drop-, or dumbbell-shaped. The glass usually appears black but on thin edges is dark green to dark brown. Tektites are named for areas where they are found in concentrations, for example, australites (Australia), indochinites (southeast Asia), and moldavites (Moldavia). Microtektites are widely distributed in soils and sedimentary rocks, but generally they can be identified only through microscopic examination. Tektites were once thought to be "moon splash," that is, hardened drops of molten liquid that were thrown out from the Moon as a result of high-energy meteorite impact. A more accepted explanation is that they form from high-energy impact melting of terrestrial material. This explanation is supported by the fact that the chemical compositions of many tektites are more like that of terrestrial than of lunar rocks. Tektites have not been reported in Iowa. The presence of the late Cretaceous Manson impact structure southwest of Fort Dodge suggests that tektites might have been produced. However, the scarcity of rocks of late Cretaceous age in Iowa makes the prospects for finding them dim.

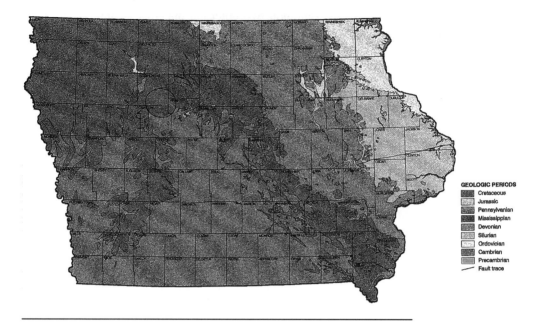

GEOLOGIC PERIODS
- Cretaceous
- Jurassic
- Pennsylvanian
- Mississippian
- Devonian
- Silurian
- Ordovician
- Cambrian
- Precambrian
- Fault trace

1.3. Geologic map of Iowa. *Geological Survey Bureau, Iowa Department of Natural Resources.*

Minerals in the Precambrian Rocks of Iowa

The oldest minerals that can be found in Iowa formed during the Precambrian, which encompasses all geologic time from the Earth's beginnings 4.6 billion years ago to about 600 million years ago (fig. 1.3; table 1.1). Except for a small exposure of quartzite in the northwest corner of the state, all Precambrian minerals in Iowa are buried under a generally thick cover of younger materials. Our knowledge of these buried minerals comes from deep drilling records and from gravity, magnetic, and seismic studies. Ancient volcanoes, eroded mountain roots, a buried continental rift, and other ancient Precambrian terrain occur in Iowa's subsurface, under several hundred meters of rock and sediment cover. Results of subsurface research reveal that large concentrations of commercially valuable minerals might lie deep beneath the land surface (Grant 1979; Ray Anderson, personal communication). Since the minerals are predominantly igneous and metamorphic, they are chiefly silicates. A good idea of the kinds of minerals that might exist beneath us comes from observation of stream gravels in Iowa. The majority of rocks and minerals found in the gravel bars of Iowa's rivers and creeks are Precambrian. They were transported here from exposures of Precambrian bedrock in northern Wisconsin, central and northern Minnesota, and southern Canada by ice sheets that advanced and retreated during the past 2 million years. Some of these rocks

Table 1.1. Geologic Column of Iowa

Era	System	Series	Episode/Supergroup	Group	Formation
Cenozoic	Quaternary	Holocene	Wisconsinan		DeForest
		Pleistocene			Peoria
					Noah Creek
					Dows
					Henry
					Sheldon Creek
					Roxana and Pisgah
			Illinoian		Glansford
					Pearl
			Pre-Illinoian	Eastern Iowa	Wolf Creek
					Alburnett
				Western Iowa	"A" glacial tills
					unnamed sediments
					"B" glacial tills
					unnamed sediments
					"C" glacial tills
	Tertiary				unnamed glacial tills
					unnamed alluvial deposits
					"Salt & Pepper Sands"
Mesozoic	Cretaceous				Niobrara
					Carlile
					Greenhorn
					Dakota
					Windrow
	Jurassic				Soldier Creek
					Fort Dodge Gypsum
					unnamed basal units

Continued

Table 1.1. *Continued*

Era	System	Series	Episode/Supergroup	Group	Formation
Paleozoic	Pennsylvanian	Virgilian		Wabaunsee	Wood Siding
					Root
					Stotler
					Pillsbury
					Zeandale
					Willard
					Emporia
					Auburn
					Bern
					Scranton
					Howard
					Severy
				Shawnee	Topeka
					Calhoun
					Deer Creek
					Tecumseh
					Lecompton
					Kanwaka
					Oread
				Douglas	Lawrence
					Cass
		Missourian		Pedee	Stranger
					South Bend
				Lansing	Stanton
					Vilas
					Plattsburg

	Kansas City	Lane
		Wyandotte
		Liberty Memorial
		Iola
		Chanute
		Dewey
		Nellie Bly
		Cherryvale
Desmoinesian	Bronson	Dennis
		Galesburg
		Swope
		Elm Branch
		Hertha
		Pleasanton
	Marmaton	Lost Branch
		Memorial
		Lenapah
		Nowata
		Altamont
		Bandera
		Pawnee
		Labette
		Stephens Forest
		Morgan School
		Mouse Creek
	Cherokee	Swede Hollow
		Verdigis
		Floris
		Spoon
		Kalo

Continued

Table 1.1. *Continued*

Era	System	Series	Episode/Supergroup	Group	Formation
	Mississippian	Atokan			Kilbourn
		Morrowan			Caseyville
		Chesterian			Pella
		Meramecian			St. Louis
					Spergen
					Warsaw
		Osagean			Keokuk
					Burlington
		Kinderhookian			"upper" Gilmore City
					"lower" Gilmore City
					Maynes Creek
				North Hill	Chapin
					Starrs Cave
					Prospect Hill
					McCraney
	Devonian	Upper Devonian		Yellow Spring	Maple Hill
					Aplington
					Sheffield
					Sweetland Creek
					Lime Creek
					Shell Rock
					Lithograph City
				upper Cedar Valley	Coralville
					Little Cedar
		Middle Devonian		lower Cedar Valley	Pinicon Ridge
					Otis
				Wapsipinicon	Spillville
					Bertram

System/Era	Series	Group	Formation
Silurian			Gower
			LaPorte City
			Scotch Grove
			Hopkinton
			Waucoma
			Blanding
			Tete des Mortes
			Mosalem
Ordovician	Upper Ordovician		Maquoketa
		"upper" Galena	Dubuque
			Wise Lake
	Middle Ordovician	"lower" Galena	Dunleith
			Decorah
			Platteville
		Ancell	Glenwood
			St. Peter
	Lower Ordovician	Prairie du Chien	Shakopee
			Oneota
Cambrian	Upper Cambrian		Jordan
			St. Lawrence
			Lone Rock
		Tunnel City	Wonewoc
			Bonne Terre
			Eau Claire
			Mt. Simon
Proterozoic	Middle Proterozoic	Carroll County	St. Augustine
			Eden
	Keweenawan		Maple River
			Westside

Continued

Table 1.1. *Continued*

Era	System	Series	Episode/Supergroup	Group	Formation
				Eischeid Farm	Brushy Creek
					Halibur
					Roselle
				Thor Volcanic	unnamed flow
					unnamed flow
					unnamed flow
				unnamed group	Finnegan Gabbro
				Mackenzie Dike Swarm	Eischeid Gabbro
				Eastern Granite-Rhyolite Terrane Pluton	Quimby Granite
					Green Island Granite
					Manson Granite
	Lower Proterozoic		Baraboo Interval Quartzites		Sioux Quartzite
					Washington County Quartzite
					Cedar Rapids Quartzite
			Central Plains Cycle Anorogenic Rocks		Hull Keratophyre
			Penokean Orogenic Rocks	Late-Orogenic Pluton	Harris Granite
					Hawarden Granite
				Camp Quest	LeMars Gneiss
					Manson Gneiss
			Penokean Taphorogenic Rocks		Matlock Banded Iron
Archean					Lyon County Gneiss
				Otter Creek Layered Mafic Complex	unnamed mafic plutonic rocks
					unnamed ultramafic plutonic rocks

Source: Geological Survey Bureau, Iowa Department of Natural Resources.

Table 1.2. Rocks Commonly Found in Iowa and Their Major Minerals

Rock Type	Mode of Occurrence	Major Minerals
Granite	glacial erratic	quartz, microcline, plagioclase
Rhyolite	glacial erratic	quartz, orthoclase, plagioclase
Gabbro	glacial erratic	plagioclase, pyroxene, ± olivine
Basalt	glacial erratic	plagioclase, pyroxene, ± olivine
Diorite	glacial erratic	plagioclase, amphibole, mica
Pegmatite	glacial erratic	quartz, microcline, mica
Quartzite	glacial erratic, bedrock	quartz
Gneiss	glacial erratic	quartz, feldspar, mica, amphibole
Slate	glacial erratic	mica, quartz, clay minerals
Chert	nodules in carbonate rock, glacial erratic	microcrystalline quartz
Sandstone	bedrock, glacial erratic	quartz
Limestone	bedrock, glacial erratic	calcite
Dolostone	bedrock, glacial erratic	dolomite
Shale	bedrock	clay minerals, microcrystalline quartz
Iron Formation	glacial erratic	chert, hematite
Ironstone	glacial erratic	limonite, clay minerals, quartz sand
Coal	bedrock	biominerals, pyrite, clay minerals
Gypsum	bedrock	gypsum, anhydrite

are similar to those brought up in deep drill cores in Iowa. These "foreign" rocks are called glacial erratics, and in addition to their presence in stream gravel, they are found as large boulders scattered across the countryside. Most of Iowa's colleges and universities have at least one large erratic on their campuses. The most abundant rock types in the erratics are granite, gneiss, basalt and chert. Table 1.2 lists common rocks found in Iowa and the major minerals they contain.

Occasionally exotic minerals are discovered in Iowa's stream gravels. Pebbles of agate are quite common. The best quality material is the so-called Lake Superior agate (LSA), which is typically reddish and exhibits concentric banding. It is believed that LSAs were originally precipitated from aqueous fluids in gas pockets in Precambrian basalt lava that spilled out onto what is now northern Wisconsin, the Upper Peninsula of Michigan, and southern Ontario. Subsequently, they weathered from the basalts that are exposed near the shores of Lake Superior and were transported by glaciers to Iowa. Rarely, a nugget of native copper is discovered. Native copper is believed to have precipitated from hydrothermal fluids that flowed through basalt and other rocks in what is now the Upper Peninsula during the Precambrian. Copper was quarried there by Native Americans and mined by Euroamericans for many years. Due to its low resistance to abrasion, native copper seldom makes it as far south as Iowa. It is easily recognized by a coating of green oxidized copper (malachite) and by its high density. In 1912 a blue sapphire was discovered in the gravels along the shore of Lake Okoboji in northwest Iowa. It was of good quality and when cut weighed one and three-eighths carats (Muilenburg 1914). The bedrock source of the sapphire is unknown, but it probably came from somewhere in Minnesota or Canada. A gem-quality green diamond was discovered in a gravel deposit at Dubuque. Its size was between that of a pea and a small hazelnut, or about five carats. Reportedly, a jeweler purchased it for $1,500 (Zeitner 1964). Like the sapphire, the diamond probably came from an unknown source in Minnesota or Canada. Other less common Precambrian minerals in Iowa's glacial erratics include hematite (from the banded iron formations of northern Minnesota and Wisconsin), black tourmaline (from pegmatites), epidote (from granite and basalt), and anorthite (from mafic igneous intrusives).

It is worth noting that glaciers brought another mineral from Minnesota and Canada—gold. It is well known that gold occurs as a minor constituent of some granitic igneous rocks. Having a relatively low melting point, it crystallized as small grains after earlier silicate minerals. Weathering of the granite released the gold, which is very stable under surface conditions. Glaciers scoured weathered granite outcrops and brought small amounts of gold along with rocks and primarily silicate minerals. These glacial deposits in Iowa have been, and continue to be, worked by creeks and rivers. Gold has been widely reported at several localities along major rivers and their tributaries, including the Des Moines, Iowa, and Volga rivers, but it does not occur in concentrations sufficient to be commercially exploitable (see chapter 4).

Minerals in Paleozoic Rocks

Minerals in Terrigenous Sandstones

Most of the bedrock exposed naturally or by human activity in Iowa is of Paleozoic age (600–250 million years ago). Every Paleozoic period, except the Permian, is represented (table 1.1). The oldest exposed Paleozoic rocks are in the northeast corner of the state, and the rocks become progressively younger toward the southwest. The youngest rocks (Virgilian) are exposed in the state's southwest corner. From Cambrian through Mississippian time Iowa was generally under the influence of shallow seas that flooded major portions of the interior of the North American continent. To the north and northeast of Iowa lay exposures of quartz-rich Precambrian crystalline rock on a low-relief topographic feature referred to as the Wisconsin Arch, which was a southern extension of the Precambrian Shield.

Streams coursing over the land surface carried sand and mud, weathered from rock outcrops, southwestward toward shallow marine basins. Sand-sized sediment was deposited along ancient shorelines, while finer clays were deposited in deeper, quieter water. Shoreline migration due to advances and retreats of the seas resulted in the wide areal distribution of quartz sand. Today extensive outcrops of sandstone, in rock units like the Cambrian Jordan and the Ordovician St. Peter sandstones, can be observed in northeast Iowa along the Upper Iowa River and along the Mississippi River and its tributaries from Dubuque north into Minnesota and Wisconsin. These same sandstones have been encountered farther south under younger rocks during water well drilling. The Jordan Sandstone is a principal supplier of groundwater in eastern Iowa. The sandstones are almost entirely quartz, but some sandstones contain appreciable amounts of feldspar. Under the microscope, occasional grains of residual minerals like garnet, tourmaline, magnetite, ilmenite, and zircon can be found. These heavy minerals were also weathered from the Precambrian igneous and metamorphic rocks of the Precambrian Shield.

In northeast Iowa sandstone contains glauconite, an iron silicate that occurs as small, dark green grains. Unlike quartz and heavy minerals, glauconite is not transported as a weathering residue. Rather, it is believed to form by chemical precipitation in a marine basin when water is shallow and oxygen availability is low. Rich glauconitic sandstone can be observed in an exposure of the Cambrian Lone Rock Formation in a small quarry on the north side of State Highway 76 a few miles west of Lansing in Allamakee County. Prolonged exposure to air causes gradual oxidation of glauconite to hematite or limonite.

From Ordovician to Pennsylvanian time little quartz sand was deposited in Iowa because source areas for quartz were distant from basins of deposition and because streams had limited energy. By the beginning of the Pennsylvanian the shallow seas had begun to withdraw, and Iowa was under the influence of freshwater streams that coursed over a generally low-relief landscape. The stream load of sand and mud was deposited in channels, in areas of overbank flooding, and in deltas, with coarser sediment generally reflecting higher stream energy conditions.

Pennsylvanian sandstones that crop out along the Des Moines River Valley in central and southern Iowa and that appear as isolated erosional remnants in the eastern part of the state contain the mineral muscovite. It occurs as small, silver-colored flakes that can be seen best as bright reflections on bedding plane surfaces. Muscovite was transported during the Pennsylvanian by streams from sites of weathering of Precambrian igneous and metamorphic rock to the north and east of Iowa. Muscovite can readily be observed at the Wyoming Hill road cut and Wildcat Den State Park along State Highway 22 east of Muscatine.

Minerals in Terrigenous Mudstones

While sand was being transported along the beds of Paleozoic streams, mud was carried in suspension in those streams. Deposition of mud occurred in offshore marine areas and in overbank areas adjacent to stream channels. The mud eventually hardened through compaction and recrystallization into mudstone and shale. The extremely small size of the individual particles in shales makes identification of minerals in them virtually impossible, even with a standard microscope. However, x-ray diffraction and electron microscopic analyses reveal that the major minerals are clay minerals (predominantly illite and kaolinite), quartz, feldspar, and calcite.

Subaerial exposure of limestones and dolostones during periods of sea regression allowed groundwater dissolution to form vug- to cavern-sized openings. The collapse of the roofs of near-surface caverns allowed streams to drain into them. Stream-carried (fluvial) sediment progressively filled the caverns, some of them entirely. Sediment filling of caverns was an important process during the early Pennsylvanian in eastern Iowa and western Illinois. Minerals in these karst-filling sediments are similar to those deposited by streams on the land surface. Paleokarst-type sediment fills are frequently exposed during limestone surface and underground mining operations in Iowa, especially in the east-central part of the state—for example, at the Buffalo, Conklin, Four County, and Robins quarries. Because conditions are ideal for preservation, excellent exposures of paleokarst-filling sediments and sedimentary structures can be seen at the Linwood Mine (Garvin 1995; Benner et al. 1997).

Minerals in Carbonate Environments

Transport of land-derived sediment into Iowa was minimal from the Ordovician through the Mississippian periods. The sea water was generally clear and shallow, with water depths ranging from subtidal to supratidal. Because Iowa was situated near the equator during the Paleozoic Era, the waters were warm and teeming with marine life. Many marine invertebrate organisms extracted dissolved mineral matter from the waters to generate skeletal material, composed primarily of calcite or aragonite, which has the same chemical composition as calcite but a different crystal structure. Some marine organisms, like sponges, secreted hard parts, probably composed of opal, which is an amorphous form of silica. Primitive fish produced teeth and bony exoskeletons composed of the phosphate mineral apatite. Apatite also comprised tiny toothlike structures known as conodonts, which are believed to be mouth parts of early chordates. Many organisms also excreted phosphatic fecal material. After the death of the organisms, predation and chemical decomposition generally destroyed all but the hard parts, which accumulated in the sediment on the floors of the shallow seas. Aragonite and opal underwent structural changes through recrystallization to form calcite and fine-grained quartz, respectively. Calcite skeletal materials and muds progressively accumulated and recrystallized to form beds of limestone. Diagenetic replacement of limestone by quartz resulted in the formation of nodules and lenses of chert, which occur abundantly in some limestone formations, such as the Silurian Scotch Grove and Hopkinton and Mississippian Burlington formations. Apatite-rich bones and teeth survived the destructive weathering and erosional processes. Conodonts are widely distributed in Paleozoic rocks, and fish teeth and head plates are common in Devonian limestones. The massive erosion caused by the overflow of the Coralville reservoir during the summer of 1993 exposed a 12-centimeter-long fragment of the head plate of a Devonian placoderm. The plate is currently on display at the Coralville Dam Visitors Center, which is maintained by the U.S. Army Corps of Engineers. Some phosphatic fecal material underwent recrystallization and selective enrichment to produce apatite-rich layers in limestone. A good example can be seen at the base of the Ordovician Maquoketa Shale near the rim of the Postville Quarry. It is important to note that some calcite in limestone might have been derived through inorganic precipitation from the tropical seas that dominated Iowa during the early and middle Paleozoic.

Following the formation of the limestone beds, some calcite was replaced by the mineral dolomite. The origin of dolomite is, and has been for many years, a subject of lively debate. Precipitation of dolomite as a primary mineral is a rare occurrence in modern marine environments. For this reason most scientists believe that the majority of the world's dolomite, including that in midcontinent Paleozoic rocks, is a product of chemical replacement of marine calcite. Replacement requires the addition of the element magnesium—the chemical formula for dolomite is $CaMg(CO_3)_2$. The timing and mechanism of replacement are not known with certainty. Replacement might have occurred soon after the accumulation of calcareous muds or long after their lithification and burial. The process of dolomitization destroys or alters the character of the original limestone, including fine-scale depositional features and fossils, which makes interpreting their history difficult. Dolomite can easily be distinguished from calcite through the careful application of dilute (10 percent) hydrochloric acid. The more reactive calcite will effervesce vigorously, whereas the less reactive dolomite effervesces sluggishly, if at all. Dolostones (carbonate rocks composed primarily of dolomite) are present in the carbonate rock record in all periods from the Cambrian through Mississippian, but they are especially abundant in rocks of Silurian and Devonian age in north-central and eastern Iowa.

Evaporite Minerals

As mentioned, Iowa was not covered continuously by seawater during the early and middle Paleozoic. Numerous episodes of sea transgression (onlap) and regression (offlap) have been documented. Major offlaps occurred during the Middle Devonian and Middle Mississippian periods. As the sea withdrew, isolated pools of water remained among exposed limestone beds. Cut off from replenishment by normal seawater and under the influence of high rates of evaporation, the water in the pools became many times more salty than normal seawater. Minerals with high solubilities in water began to precipitate from these hypersaline waters and gradually accumulated as beds of gypsum. Burial, with a corresponding increase in pressure, resulted in the dehydration of gypsum and its replacement by anhydrite. Subsequent diagenetic modifications of these evaporite deposits include rehydration of anhydrite to form a new generation of gypsum and the precipitation of celestite on cavity walls. Much of this evaporitic material was later dissolved by groundwater, and the overlying limestone beds collapsed into the empty spaces thus created. Regional removal of evaporite minerals is evidenced by widespread brecciation of the middle Devonian Davenport Member of the Pinicon Ridge Formation. However, in central and southeast Iowa undissolved evaporites have been observed in drill cores. At the United States Gypsum's Sperry Mine, gypsum and anhydrite from the same stratigraphic interval have been mined for many years (Dorheim et al. 1972; Sendlein 1973).

Minerals in Coal

Coal-bearing strata underlie the southwestern third of Iowa and comprise the northernmost part of the Western Interior Coal Basin (fig. 1.4). These beds center on the Forest City Basin, where Pennsylvanian rocks reach a maximum thickness of 500 meters. The outcrop area lies along the northeastern edge of this basin. Isolated occurrences of coal-bearing Pennsylvanian rocks are found in east-central Iowa, including the outcrops at Wyoming Hill near Muscatine. These coal beds lie along the western edge of the Eastern Interior Coal Basin, which centers in Illinois (Hinds 1909; Van Dorpe and Howes 1986).

The process of coal formation began approximately 300 million years ago. Changes in land/sea relationships, mediated by climate-driven global changes in sea level and tectonic activity in the eastern and southern regions of the North American continent, resulted in the withdrawal of epicontinental seas and in the generation of low-relief coastal swamps. Related modern-day environments include the Florida Everglades and the Louisiana Bayou. Iowa's climate during the Pennsylvanian was hot and humid. Stagnant waters and abundant tropical vegetation favored the accumulation of partially decomposed plant material and its conversion to peat. Repeated cycles of sea transgression and regression produced alternating beds of terrigenous clastic sedimentary rocks, peat, and thin beds of limestone (Van Dorpe and Howes 1986). At least forty transgression-

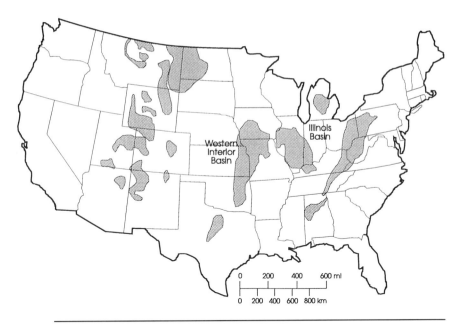

1.4. Locations of major coal occurrences in the United States, emphasizing the Western Interior and Illinois coal basins. *After Craig et al. (1988).*

regression (T-R) cycles during the Pennsylvanian have been identified in Iowa (Heckel 1986). Pressure and heat of burial gradually transformed the peat into lignite and the lignite into bituminous coal. The coalification process generated a group of minerals that are termed "biogenic" because they are formed from chemical components originating in the coal and because their formation is believed to have been mediated by coal-associated anaerobic bacteria. The most prominent of these minerals is pyrite, which occurs as fine-grained disseminations in the coal and black shales that typically overlie coal beds. Pyrite probably results from conversion of preexisting iron monosulfides like mackinawaite (Fe_9S_8) and greigite (Fe_3S_4) (Blatt et al. 1980). These pyritic disseminations, which are partly responsible for the high sulfur content of Iowa coal, are redistributed locally to form concretionary masses and fracture fillings. Other sulfides, including marcasite and sphalerite, have also been reported from coal beds in Iowa (Hatch et al. 1976; Garvin, unpublished data).

Minerals in Mesozoic Rocks

Jurassic Evaporite Minerals

The seas withdrew from Iowa during the late Pennsylvanian and did not return until the late Jurassic Period. The Interior Seaway was centered well to the west of Iowa, but embayments and lagoons extended as far east as

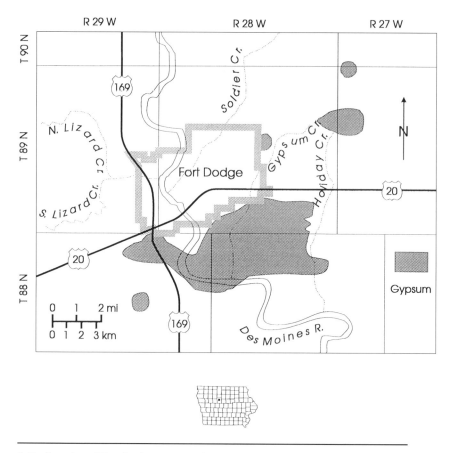

1.5. Location of Fort Dodge gypsum deposits. *After Dasenbrock (1984).*

central, and perhaps eastern, Iowa. The presence of isolated bodies of water and the arid to semiarid climate of the late Jurassic in Iowa provided favorable conditions for the formation of evaporite deposits. At Fort Dodge the deposits are virtually all gypsum and range from 1 to 9 meters (average 5 meters) thick (fig. 1.5). Chemical analyses of seawater demonstrate that, if a standing column of seawater evaporates to dryness, about 85 percent of the mineral deposit will be halite and only 2 percent gypsum; yet there is no halite at Fort Dodge. The absence of halite might be a consequence of the conditions of primary deposition in the evaporite basin. The general idea is that at the time of gypsum precipitation, evaporite brines were still undersaturated with respect to halite. Remaining brines, being dense, might have sunk into the floor of the basin or might have flowed out of the basin beneath an influx of new seawater (Bard 1982; Dasenbrock 1984) (fig. 1.6). Repeated influxes of normal seawater followed by extended periods of evaporation are believed to account for large thicknesses of gypsum. Alternatively, halite might have been deposited originally but, owing to its very high solubility in water, was later removed by ground-water dissolution.

Fine-scale laminations, which are characteristic of Fort Dodge gypsum beds, may be a result of seasonal changes in the evaporite basin (Bard 1982). Certain structural features in the gypsum beds provide evidence that some gypsum may have been converted to anhydrite, a result of de-hydration due to the pressures of sediment burial (Dasenbrock 1984). The absence of anhydrite in the deposits today is attributed to subsequent ero-sion of the overburden, which allowed anhydrite to rehydrate to gypsum. The presence of fibrous gypsum veins is believed to be a result of diage-netic dissolution and recrystallization of gypsum along bedding plane frac-tures (Bard 1982; Dasenbrock 1984; Hayes 1986). Following the Jurassic Period, much of the gypsum apparently was removed by erosion, but iso-lated deposits remain in Webster County in the vicinity of Fort Dodge (Cody et al. 1996).

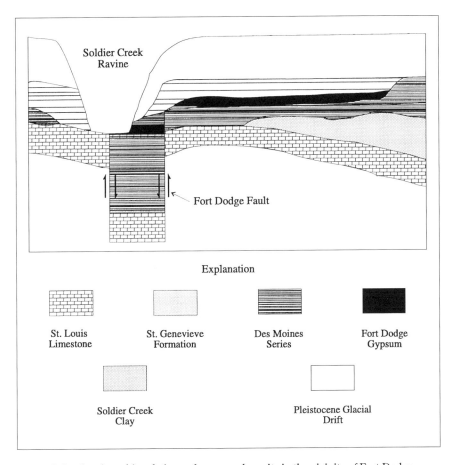

1.6. Stratigraphic relations of gypsum deposits in the vicinity of Fort Dodge. *After Bard (1982).*

The minerals that formed during the Cretaceous Period in Iowa are products of both marine and nonmarine depositional environments, and they reflect early coastal deltaic and fluvial, and later marine shoreline and coastal lagoon, conditions (Tester 1929). The minerals deposited under marine conditions were primarily terrigenous quartz and clay minerals and marine calcite. The fluvial deposits include quartz, clay minerals, and lithic fragments, as well as minerals associated with thin beds of lignitic coal. Pyrite is locally abundant as concretionary masses in shale and sandstone. Gypsum, found as scattered crystals and as narrow vein fillings in shale and siltstone, likely results from the diagenetic decomposition of pyrite. Chalcedony (petrified wood) has been reported from the Dakota Formation in northwest Iowa and has also been found in gravels along the Cedar River drainage in eastern Iowa (Horick 1974; Garvin, unpublished data). The latter occurrence may have been derived by glacio-fluvial transport from isolated exposures of Cretaceous rocks in Floyd County (Ray Anderson, personal communication).

Iron Ores of Allamakee County

In extreme northeast Iowa are scattered accumulations of hydrous iron oxide known as the Iron Hill Deposit. The largest concentrations occur about 5 kilometers northeast of Waukon. They appear to be related to much larger deposits in southeast Minnesota, which are discussed in detail by Bleifuss (1972). Historically they have been assigned to the Cretaceous Windrow Formation, though Bleifuss believes that they formed during the early Tertiary Period. The ores consist primarily of goethite, which occurs as hard, dark brown nodular masses exhibiting a fibrous habit, as coarse cellular masses, or as soft, yellow-brown earthy masses. Subordinate hematite is also observed. In Minnesota small amounts of pyrite and the clay mineral illite have been reported, as have localized masses of the iron carbonate siderite. The iron-bearing material rests on dolostones of the Ordovician Galena Formation (in Minnesota it is also found on rocks of the Devonian Cedar Valley Group). The ore contains angular chert nodules, which are variably replaced and cemented by iron oxide. Also present are limestone fragments of various sizes up to 3 meters across. Goethite locally exhibits stalactitic/stalagmitic habit (Howell 1915).

The preceding information suggests that the iron ores are products of the weathering of underlying dolostone bedrock. However, chemical similarities between the iron ores and the siderite in Minnesota led Bleifuss to conclude that siderite, not dolomite, was the protore (fig. 1.7). Siderite is believed to have been deposited in a near-shore marine environment as a primary mineral in strata of the Cedar Valley Group. The sideritic

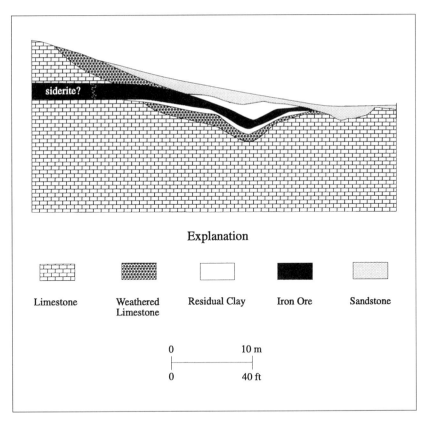

siderite?

Explanation

Limestone	Weathered Limestone	Residual Clay	Iron Ore	Sandstone

0   10 m

0   40 ft

1.7. Stratigraphic relations of Cretaceous iron deposits in southeast Minnesota. *After Bleifus (1972).*

beds were buried by younger Devonian rocks, where they remained until erosion exposed them during the late Cretaceous or early Tertiary. Oxidation of the siderite under temperate climatic conditions (which is evidenced by the presence of illite) is believed to have formed the iron oxides (Bleifuss 1972).

Epigenetic Minerals of Uncertain Age

Most of the collectible minerals in Iowa come from deposits whose ages are not known with certainty. They occur as open-space linings and fillings in, and as chemical replacements of, the rocks that enclose them. They are clearly younger than their hosts but older than the glacial deposits of the Pleistocene. Because they are younger than the enclosing rock, they are referred to as epigenetic minerals. They can be found in rocks of all ages from Cambrian through Pennsylvanian, and they are especially abundant in the eastern half of the state. They occur as linings and fillings in caverns, vugs, fractures, and geodes; as breccia cements; and as concretionary and other replacements in limestone, dolostone, and shale.

Table 1.3. Minerals Occurring in Epigenetic Deposits of Uncertain Age in Eastern Iowa

Sulfides	Carbonates	Halides	Sulfates	Silicates
Chalcopyrite $CuFeS_2$	ankerite $CaFe(CO_3)_2$	fluorite CaF_2	barite $BaSO_4$	chalcedony SiO_2
Galena PbS	calcite $CaCO_3$		celestite $SrSO_4$	chert SiO_2
Marcasite FeS_2	dolomite $CaMg(CO_3)_2$		gypsum $CaSO_4 \cdot 2H_2O$	quartz SiO_2
Millerite NiS	siderite $FeCO_3$			
Pyrite FeS_2				
Sphalerite ZnS				
Wurtzite ZnS				

Note: This table does not include minerals resulting from chemical weathering of primary minerals, such as hematite, limonite, anglesite, and smithsonite.

Table 1.3 lists the common carbonate-hosted epigenetic minerals that have been reported in eastern Iowa. The majority of the localities are quarries and mines (limestone or lead). For convenience of comparison they are listed in table 1.3 by anion chemical group. Of course, no single deposit contains all of the minerals, but twelve minerals are found in geodes, and ten occur in a single quarry (Pint's Quarry).

These epigenetic mineral deposits are located along the western margin of the formerly commercial Upper Mississippi Valley Zinc-Lead District (UMV) (fig. 1.8). Mines at Dubuque, Guttenberg, Lansing, and Mineral Creek are considered part of the district (Heyl et al. 1959). These world-famous deposits have been the objects of intensive study ever since they were described by David Dale Owen in 1840. Detailed descriptions of the commercial mines in the UMV, including those in Iowa, are given by Heyl et al. (1959). Lead ore in Iowa was contained primarily in solution-enlarged high-angle fractures in rocks of the Ordovician Galena Group (table 1.1). Referred to as crevices and gash veins by early miners, these fractures have a generally eastward trend. The major primary (i.e., unoxidized) minerals are galena, sphalerite, marcasite, pyrite, calcite, and barite. Quartz and dolomite replaced host-rock minerals in advance of the main mineralizing events. The sulfide-bearing deposits that were encountered during early stages of mining, being near the surface, were highly oxidized. Iron sulfides have been replaced by hematite and limonite, sphalerite by smithsonite and hemimorphite, and galena by cerussite and anglesite. The lead ores at Lansing are also of the crevice type, except that the controlling fissure is oriented north-south. The primary mineral at Lansing is galena; minor amounts of pyrite, marcasite, and calcite have also been documented (Garvin, unpublished data). At Mineral Creek the ores are largely breccia cements. The brecciation occurs near the crest of an east-west-trending anticline (Garvin 1982). The primary minerals at Mineral

Creek are essentially the same as those in crevice deposits in the main district, except that barite is absent. Both the Lansing and Mineral Creek deposits are contained in Ordovician Oneota Dolostone, and in both the minerals are highly oxidized. At Dubuque crevice deposits give way with increasing depth to low-angle and horizontal fracture–filling deposits, known as pitches and flats (fig. 1.9).

Many theories of origin have been advanced since the discovery of UMV mineral deposits. It is generally agreed that the deposits are epi-thermal, which means that they were precipitated from hydrothermal fluids at relatively low temperatures (50–225°C) and shallow depths. The source of heat for the fluids is a matter of debate. These heated fluids

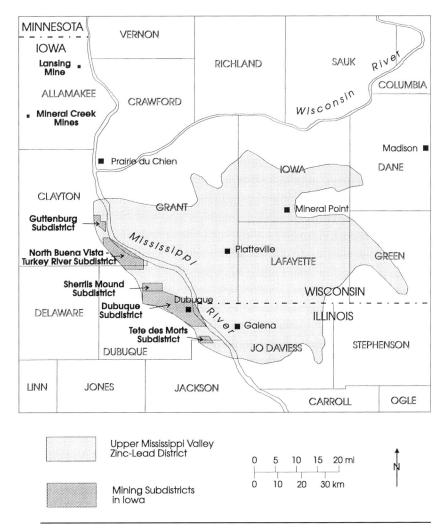

1.8. Locations of lead and zinc mining areas in northeast Iowa in relation to the Upper Mississippi Valley Zinc-Lead District. *After Heyl et al. (1959).*

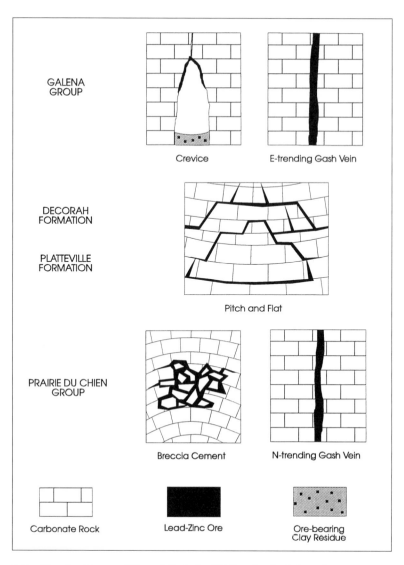

1.9. Structural types of Upper Mississippi Valley zinc-lead deposits.

(which were probably sedimentary pore waters that in late stages of mineralization were modified by meteoric water) dissolved mineral-forming chemicals from surrounding marine sedimentary rocks as they moved laterally and vertically along fractures and other avenues of flow. Cooling of the fluids and their chemical reactions with carbonate host rocks are believed to be the primary mechanisms responsible for the precipitation of the minerals. Precipitation typically began with early iron sulfide deposition (pyrite and marcasite) on the walls of open fractures and in voids in breccia. Where fractures remained open, galena and sphalerite were deposited next upon the iron sulfide minerals. In large crevices galena formed crusts a half meter or more thick. As fluids continued to cool, calcite and, locally, barite precipitated. Rubidium-strontium isotopic dating of UMV sphalerite indicates that mineralization occurred about 270 million years ago (Brannon et al. 1992).

Minor sulfide-bearing mineral occurrences are widely distributed in Iowa, especially in the eastern part. Figure 1.10 illustrates the distributions of some epigenetic minerals in eastern Iowa. Pyrite, marcasite, and sphalerite are present in virtually all of these deposits (refer to the descriptions

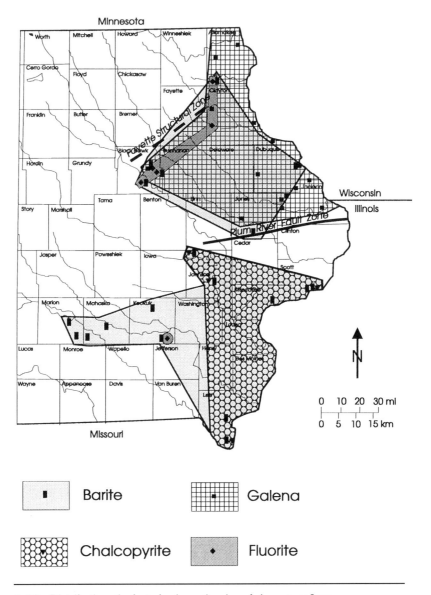

1.10. Distribution of selected epigenetic minerals in eastern Iowa.

of individual mineral deposits in chapter 3). Fluorite and galena are confined to northeast Iowa, whereas millerite and chalcopyrite occur in the southeast. These minor epigenetic mineral occurrences have several characteristics in common with UMV deposits. First, all Iowa minerals except fluorite occur in UMV deposits. Second, both types of deposits are hosted by Paleozoic carbonate rocks. Third, structural controls on mineral deposition are broadly similar (e.g., bedding plane and high-angle fractures, breccia voids, caverns, and vugs). More detailed comparisons reveal that some Iowa mineral deposits are more like UMV deposits than others. Comparisons are based chiefly on mineralogy, paragenesis, and stable isotope geochemistry. Deposits most like UMV deposits include those at the Cedar Rapids, Conklin (part), and Ferguson quarries (Garvin 1984b; Garvin et al. 1987; Garvin and Ludvigson 1988, 1993; Spry and Kutz 1988; Kutz and Spry 1989). The similarities suggest that these Iowa deposits and those at UMV might have formed under similar environmental conditions, perhaps as part of a regional-scale mineral-forming event (table 1.4).

Mineral deposits that show marked differences in paragenesis and geochemical character with those from UMV are thought to result from processes that operated on a local scale (table 1.4). These deposits include those at the Conklin (part), Fairbank, Four County, and Pint's quarries (Anderson and Garvin 1984; Garvin et al. 1987; Garvin and Ludvigson 1988, 1993; Spry and Kutz 1988; Kutz and Spry 1989). All chemicals essential to the formation of these minerals can be obtained in adequate quantity by leaching surrounding rocks. The sources of the mineralizing fluids are not known.

Stable isotopic compositions of epigenetic mineral deposits at the Buffalo, Keswick, Moscow, Ollie, Peske, Postville, and Robins quarries, at the Linwood and Waterloo South mines, at the Volga exposure, in the geode deposits in southeast Iowa, and in deposits associated with the coal beds of south-central Iowa have not been determined. Therefore, adequate comparisons with UMV deposits cannot be made. Regardless whether any of these minor sulfide-bearing deposits in Iowa are or are not cogenetic with UMV deposits, considering their individual physical and chemical characteristics, each mineral deposit is the product of a unique process (Brown 1967). Given the fact that virtually all of these epigenetic minerals are known to be stable precipitates at surface temperatures, some of these minerals might have precipitated from unheated groundwater.

Several epigenetic deposits are spatially associated with pre-Pennsylvanian caves, which are abundant in eastern Iowa and adjacent western Illinois. The caves, their sediment fills, and the minerals lining their surfaces are especially well preserved at the Linwood Mine, a room and pillar limestone mine, near Davenport. The walls and pillars afford an excellent opportunity to study this mineral-forming environment (Garvin and Crawford

1992; Garvin 1995). The mineralogy of cave-filling stream sediments was previously discussed. Where organic-rich mudstones occur in the cave sediment, mineral sulfides are invariably present. Nodular and disseminated pyrite and marcasite are found within the mudstone, and these minerals, along with sphalerite, chalcopyrite, calcite, and barite, occur as scattered crystals and crusts on cave walls and on breakdown blocks resting on cave floors.

Phantom calcite crystals, such as those occurring at the Linwood Mine and the Moscow and Buffalo quarries, provide evidence for several stages of mineral growth. Rhombohedral forms, overgrown by scalenohedral forms, or vice versa, are easily recognized because of microscopic pyrite and marcasite "dust" on interior growth surfaces (fig. 1.11).

The presence of sulfide minerals in virtually all of the epigenetic mineral deposits indicates reducing conditions during mineral deposition, meaning that oxygen levels were low. Such conditions exist below the groundwater table, especially where there is abundant partially decayed organic matter and/or previously formed iron sulfide. Thus, mineral-lined fractures and cavities were completely filled with water at the time of sulfide mineral precipitation. Mineral deposition may have resulted from cooling of hydrothermal fluids or from changes in the chemistry of the depositing fluids by reaction with enclosing rocks or through mixing of waters of different compositions. After formation of the minerals, lowering of the groundwater table changed the subsurface environment from reducing (phreatic) to oxidizing (vadose). Sulfide minerals, which are unstable in the presence of moist air, were gradually coated with oxides. Pyrite and marcasite oxidized to goethite, hematite, and lepidocrocite; chalcopyrite to malachite and goethite; sphalerite to smithsonite and hemimorphite; and galena to anglesite and cerussite. In some cases oxidation was complete, but the replacement process preserved the outer form of the original sulfide mineral, a process known as pseudomorphism, which enables identification of the original mineral. In other cases unaltered cores of original sulfide minerals still remain, for example, at the Mineral Creek Mines (Garvin 1982). At the Dubuque Mines oxidation of iron sulfides and sphalerite was virtually complete in the upper parts of the deposits, while galena, which is more resistant to the effects of atmospheric oxidation, survived, armored by a thin coating of anglesite or cerussite. Oxidation of iron sulfide minerals in coal and black shale frequently produced gypsum.

Origin of Iowa Geodes

Although the minerals found in Iowa geodes are, like other epigenetic minerals, considered to be products of precipitation from aqueous fluids, they deserve special mention because they are so well known and because the process of geode formation is especially interesting. The best-known

Table 1.4. Comparison of Epigenetic Mineral Deposits in Eastern Iowa with Main District Upper Mississippi Valley (UMV) Zinc-Lead Deposits

Deposit	Primary Minerals	Paragenesis	Primary Mode(s) of Occurrence	Host Rock Age	Sulfide Sulfur Isotopic Comparison (δ^{34}S)
Main District UMV	ba, cc, cp, do, gn, mc, py, sp[a]	early sulfide–late cc	pitch flat, crevice, cavity linings, breccia cement	Ordovician	5 to 28[b]
Cedar Rapids Quarry	cc, mc, py, sp	early sulfide–late cc	cavity lining, fracture filling, breccia cement	Silurian	paleokarst: 12.9 to 21.4 fracture fill: 1.0 to 30.3[c]
Conklin Quarry	ba, cc, cp, mc, ml, py, sp	type I: early sulfide–late cc; type II: early cc–late sulfide	cavity lining, fracture filling	Devonian	type I: −33.8 to 15.7 type II: −25.1 to 14.3[c]
Fairbank Quarry	cc, mc, py	early sulfide–late cc	cavity lining, fracture filling	Devonian	−0.3 to −19.6[d]
Ferguson Quarry	cc, mc, py	early sulfide–late cc	cavity lining, fracture filling	Mississippian	2.8 to 33.1[d]
Four County Quarry	ba, cc, cp, mc, ml, py, sp	early cc–late sulfide	cavity lining, fracture filling	Devonian	−30.9 to 18.4[c]
Keswick Quarry	ba, cc, mc, qz, py, sp	early cc–late sulfide	cavity lining	Devonian	
Lafarge Quarry	ba, cc, do, mc, py, sp	early sulfide–late cc	cavity lining	Devonian	

Lansing Mine	cc, gn, mc, py	early sulfide–late cc	crevice	Ordovician–Cambrian	−16.1 to 5.1[e]
Linwood Mine	ba, cc, cp, mc, py, sp	early sulfide–late cc	cavity lining, breccia cement	Devonian	
Mineral Creek Mines	cc, gn, mc, py, sp	early sulfide–late cc	breccia cement	Ordovician	3.9 to 15.2[e]
Moscow Quarry	cc, do, mc, py, sp	early sulfide–late cc	cavity lining	Devonian	
Ollie Quarry	ba, cc, mc, ml, py, qz, sp	early cc–late sulfide	cavity lining	Devonian	
Peske Quarry	ba, cc, fl, mc, py	early cc–late sulfide	cavity lining	Devonian	
Pint's Quarry	ba, cc, fl, gn, mc, py, qz, sp	early cc–late sulfide	cavity lining	Devonian	−15.9 to −9.9[e]
Postville Quarry	ba, cc, fl, gp, mc, py, sp	early cc–late sulfide	cavity lining	Ordovician	
Robins Quarry	cc, mc, py, sp	early sulfide–late cc	cavity lining, fracture filling	Devonian	
Volga	cc, fl, py, sp	early sp–late cc	cavity lining	Ordovician	
Waterloo South Mine	ba, cc, fl, mc, py, sp	early cc–late sulfide	cavity lining	Devonian	

[a] ba = barite, cc = calcite, cp = chalcopyrite, do = dolomite, fl = fluorite, gn = galena, gp = gypsum, mc = marcasite, ml = millerite, py = pyrite, qz = quartz, sp = sphalerite.

[b] Source: McLimans (1977) (for pitch/flat-type deposits).

[c] Source: Garvin and Ludvigson (1993).

[d] Source: Kutz and Spry (1989).

[e] Source: Garvin et al. (1987).

1.11. Calcite phantom crystals. Early rhombohedral crystal inside later scalenohedral crystal. Growth zones are accentuated by microcrystalline marcasite "dust." Linwood Mine, Scott County. Specimen is 7 centimeters long. *Anderson Museum, Cornell College.*

definitive works on the origin of Iowa geodes are by Van Tuyl (1922), Hayes (1964), and Chowns and Elkins (1974). Geodes are sedimentary nodules consisting of a chalcedony rind and often a drusy quartz interior. They are spheroidal masses that exhibit varying degrees of flattening parallel to the bedding planes of the enclosing carbonate rocks. Some appear to have been crushed, with fragments cemented together by mineral matter. Not all geodes have hollow centers, and not all contain well-terminated quartz crystals, but all display an outer rind of chalcedony, which distinguishes them from vugs.

The spheroidal shape is believed to result from the fact that the space occupied by a geode was formerly occupied by another object—a concretion. A concretion is a more or less spheroidal mass (although it may take a more irregular shape) of mineral matter that appears to have grown outward from a center. Sometimes the center is defined by a nucleus of different material, such as a fossil or a grain of sediment. Concretions grow by cementing the sediment of the enclosing rock or by replacing it. They may form in sandstones, shales, and limestones and may be composed of calcite, quartz, pyrite, gypsum, barite, or other minerals. The concretions related to Iowa geodes were probably calcite (Hayes 1964; Maliva 1987) or anhydrite (Chowns and Elkins 1974; Witzke 1987).

The host rocks enclosing Iowa geodes are shaly dolostones and dolomitic mudstones of the Mississippian Warsaw and Keokuk formations (table 1.1). According to Hayes (1964) and Chowns and Elkins (1974), concretions formed in these units soon after the lime muds were deposited. The most convincing evidence is that sedimentary layers are draped over concretions and geodes, indicating that the overlying sediments were still unconsolidated and thus that the concretions were near the sediment-water surface at the time they formed. Unreplaced concretions occur locally within the Warsaw.

The process of Iowa geode formation began with silica replacing, volume for volume, calcite or anhydrite at the margin of the earlier-formed concretion. This replacement produced the chalcedony rind, present in all geodes, that protects geode interiors from destruction by weathering and erosion. Evidence of pressure-induced outward expansion of silica rinds during replacement of concretion margins (Pettijohn 1957) is lacking in Iowa geodes. Concurrent with the replacement of the concretion rim was the recrystallization of its core and its replacement by chalcedony and microcrystalline quartz. Next, the recrystallized core was dissolved, possibly by acidic solutions. Minerals were later deposited in the hollow core by fluids that entered through micro- and macro-fractures in the rinds. Anhedral and euhedral quartz is the most common cavity-lining mineral, and the most attractive geodes exhibit lustrous, well-terminated quartz crystals. Other minerals were also deposited as cavity linings and fillings, including calcite, dolomite, pyrite, marcasite, chalcopyrite, sphalerite, barite, and gypsum. Some geodes, especially those hosted by dolomitic mudstones, were crushed due to the pressure of overlying beds and the low strength of the enclosing rock. Fragments of crushed rinds were subsequently cemented together by the previously named minerals (Sinotte 1969) (fig. 1.12).

The widespread presence of pyrite in geodes and in their enclosing rocks indicates that mineral precipitation took place under phreatic conditions. The presence of oxide coatings on many of the sulfide minerals indicates that, after primary mineral formation, the water table dropped, changing the geode environment from phreatic to vadose. As stated earlier, similarities between minerals in Iowa geodes and those in UMV deposits suggest that some geode minerals may be hydrothermal.

Stalactitic Goethite

Another rather curious mineral deposit in Iowa deserves mention. Near the town of Andrew in Jackson County, masses of goethite are found in isolated locations along stream drainages. They appear as large flat slabs up to 30 centimeters across, from which project knobby or cylindrical masses as much as 15 centimeters long and 8 centimeters in diameter, and as irregularly shaped clusters. Axes of knobs projecting from slabs are in parallel alignment. Cylinders less than 1 centimeter across and at least

1. Deposition of
   host shale

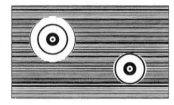

2. Formation of
   anhydrite nodules

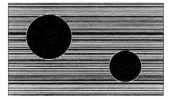

3. Dissolution of
   anhydrite nodules

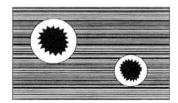

4. Precipitation of
   geode minerals

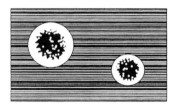

5. Oxidation of
   sulfide minerals

1.12. Sequence of formation of Iowa geodes.

4 centimeters long are aligned in parallel clusters. Close inspection of the goethite reveals that it is pseudomorphic after pyrite and marcasite. Original pyrite cubes were as much as 3 centimeters across. The cores of some cylindrical masses contain unaltered pyrite (Steven Barnett, personal communication). The goethite occurs as float in stream gravels, which also contain abundant irregular masses of iron-stained chert. Although none of the goethite has been found in place, its crystal habit can be described as stalactitic, suggesting that it formed on ceilings and floors of open cavities, at least some of which must have been quite large. The age and manner of formation of the original iron sulfide stalactites are not known. It is generally assumed that stalactites and other dripstone formations precipitate from the dripping action of water in air-filled cavities. Yet pyrite, which requires reducing conditions for stability, seems to require a phreatic environment, that is, water-filled cavities. This apparent contradiction has not

been resolved. Stalactitic iron sulfide is known elsewhere in the area. Small (<3 centimeters long) pyrite stalactites have been observed in place in small cavities in the Cedar Rapids Member of the Middle Devonian Otis Formation at the Linwood Mine (Garvin 1995). Stalactitic masses of pyrite, reportedly the size of baseball bats, projecting vertically from the ceilings of large cavities in Silurian carbonate rocks have been reported in the Midway Quarry at Joslin (near Moline, Illinois). They were collected years ago by the late Art Blocher of Amboy, Illinois, and for years were stored in piles in his yard. The iron sulfide minerals are partly overgrown with clusters of sphalerite crystals.

Quaternary and Recent Minerals

Cave Deposits

Mineral deposits of Quaternary and Recent age have been reported from several caves in northern and eastern Iowa. Probably best known are those deposits in Cold Water Cave, which was discovered near Decorah in 1969. Cave minerals (speleothems) are believed to form when water containing dissolved $CaCO_3$ enters the cave along fractures in the carbonate rock ceiling. Change in the pressure of carbon dioxide in the cave atmosphere causes precipitation of calcite dripstone formations. While pure calcite is white, the dripstones often exhibit red, brown, or black coloration due to the presence of iron and manganese impurities. Caves in eastern Iowa are believed to have developed following the last major ice incursion into eastern Iowa. The layers in speleothems record changes in chemical and climatic conditions during growth, changes that can be interpreted through studies of stable isotopes of carbon and oxygen that are contained in the calcite.

Rates of growth of speleothemic calcite are generally very slow. At Carlsbad Caverns in New Mexico, for example, the average growth rate is likened to adding a coat of paint (<1 millimeter) every hundred years. More rapid growth rates may occur locally, as at the Linwood Mine, where calcite dripstone is currently forming through the action of water cascading into a mine opening (room) through large fissures in the ceiling. The rate of calcite precipitation is estimated to be 1 centimeter in fifteen years (the length of time that portion of the mine has been opened), which extrapolates to more than 6 centimeters in a hundred years (Garvin 1995).

Sulfate Bloom on Carbonate Rock, Organic-rich Shale, and Coal

As was discussed previously, iron sulfide minerals are common as disseminations of microscopic crystals in organic-rich shale, coal, and some carbonate rocks. Having originally formed under chemically reducing conditions, they can decompose rapidly in the presence of moist, oxygen-rich air. The ultimate products of decomposition are typically hematite and goethite and other hydrous iron oxides. A byproduct of decomposition is sulfuric acid which, when present in sufficient quantity, acidifies waters draining through the decomposing material. This acid drainage constitutes a serious environmental hazard, adversely affecting aquatic plant and animal life.

The initial breakdown of iron sulfide produces soluble sulfate (SO_4^{2-}). During periods of evaporation (dry seasons) the sulfate may quickly combine with calcium, magnesium, or iron, which are relatively abundant in surrounding shale and carbonate rock, and may precipitate as an efflorescence, or bloom, on rock surfaces. Epsomite is ubiquitous as a white coating on dolostone outcrops, quarry walls, and building stones. It is easily rubbed off with the hand and is quickly washed away by rainfall, only to bloom again during the next period of evaporation. On outcrops of organic-rich shale and coal, blooms of iron sulfate minerals are common. They appear most often as white, yellow, or sometimes pink powdery coatings in overhangs and other areas where they are protected from dissolution by rainwater. Most common among the iron sulfates are halotrichite, rozenite, szmolnokite, and melanterite. These minerals have been observed at surface exposures of coal in mining areas and rock outcrops in south-central and eastern Iowa. A good example is found in a narrow seam of coal exposed in a large road cut at Wyoming Hill, near Muscatine. Nearly transparent fibrous blue masses of melanterite were collected from an exposure of karst-filling black shale in the Linwood Mine (Garvin 1995). The mineral appears to have formed after exposure of the shale to air during mining. Upon removal from the mine, melanterite quickly (within a day or so) decomposes to white, powdery szmolnokite. The identities of both minerals were confirmed by x-ray powder diffraction. Table 1.5 compares the physical properties of the more common hydrated sulfates as an aid to their identification. Like epsomite, iron sulfates in general are highly soluble in water; thus they are quickly dissolved during rainfall, only to reappear during subsequent evaporation.

The mineral vivianite (a hydrated iron phosphate) occurs in iron- and clay-rich glacial materials overlying Paleozoic bedrock near the Saylorville Dam and at Ledges State Park. It is associated with a white efflorescent mineral (an iron sulfate or epsomite). Vivianite is bright blue, but the color quickly fades and the luster dulls on exposure to light. The mineral is believed to be an alteration product of preexisting iron carbonates or sulfides.

Extraterrestrial Minerals

Although Iowa is not as well known as Arizona and Texas as a place to find meteorites, falls and finds are known from several localities (fig. 1.13, table 1.6). Extensive searches for meteorites have been made in Texas,

Table 1.5. Properties and Modes of Occurrence of Iron Sulfate Minerals

Name	Chemical Formula	Color[a]	Habit	Mode of Occurrence
Coquimbite	$Fe_2(SO_4)_3 \cdot 9H_2O$	**violet**; also yellowish, greenish; colorless	short prismatic; massive; granular	with other secondary sulfates
Ferrohexahydrite[b]	$FeSO_4 \cdot 6H_2O$	colorless, white		
Halotrichite	$FeAl_2(SO_4)_4 \cdot 22H_2O$	**colorless-white**; yellowish-greenish	acicular; tufted; spheroidal	efflorescence from alteration of pyrite; recent deposit in mine workings
Kornelite	$Fe_2(SO_4)_3 \cdot 15H_2O$	pale rose–pink to violet	acicular; crusts; tufted; globular	alteration of pyrite
Lausenite[c]	$Fe_2(SO_4)_3 \cdot 6H_2O$	white	lumpy aggregate of minute fibers	with other sulfates as result of fire in United Verde Mine, Jerome, Ariz.
Melanterite	$FeSO_4 \cdot 7H_2O$	**green** to blue; bluer with more Cu	fibrous; stalactitic	oxidation of pyrite and marcasite; efflorescence on walls in mines and sheltered outcrops
Quenstedite	$Fe_2(SO_4)_3 \cdot 10H_2O$	**pale violet** to red-violet	aggregate of minute crystals	found with coquimbite and copiapite
Roemerite	$FeFe_2(SO_4)_4 \cdot 14H_2O$	**rust brown** to yellow	tabular; crusts; granular; stalactitic	from oxidation of pyrite; found with other Fe sulfates
Rozenite[b]	$FeSO_4 \cdot 4H_2O$	colorless, white		
Siderotil	$FeSO_4 \cdot 5H_2O$	**white** to **yellow**; pale green-white	fibrous crusts; needles	found with melanterite; possible alteration product of melanterite
Szmolnokite	$FeSO_4 \cdot H_2O$	**yellow**, red brown; also blue, colorless	tabular; globular	associated with pyrite and secondary sulfates; deposited from high acid and concentrated solutions

[a] Bold type indicates most common color.

[b] Sources: Joint Committee for Powder Diffraction Standards book file; all other data from Palache, Berman, and Frondel (1951).

[c] Not listed in Joint Committee for Powder Diffraction Standards book file.

Kansas, and Nebraska with very fruitful results. To my knowledge, searches of this kind have not been made in Iowa. Despite the fact that, in comparison to the climate of semiarid western states, Iowa's climate is conducive to meteorite weathering, the prospects for meteorite discovery here ought to be good, since there is much open, rock-free land. The average rate of fall of meteors having masses greater than 500 grams (the minimum size considered necessary for survival from atmospheric heating) is estimated

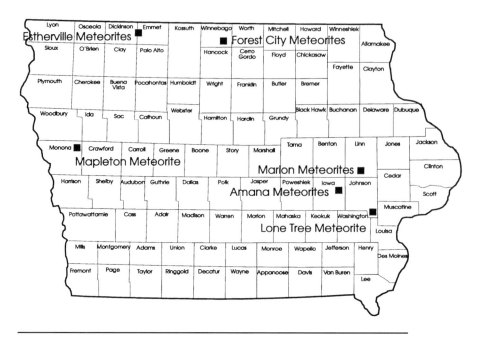

1.13. Locations of reported meteorite impacts in Iowa.

Table 1.6. Recorded Meteorite Discoveries in Iowa

Name	County	Date Discovered	Type	Weight (kg)
Marion	Linn	Fell Feb. 25, 1847	stone	28
Amana (Homestead)	Iowa	Fell Feb. 12, 1875	stone	227
Estherville	Emmett	Fell May 10, 1879	stony iron	338
Forest City	Winnebago	Fell May 2, 1890	stone	122
Marengo	Iowa	Fell March 27, 1894	stone	unknown
Mapleton	Monona	Found June 17, 1939	iron	49
Lone Tree[a]	Johnson	Found 1971	stony iron	21

Sources: Hinrichs (1905); Nininger and Nininger (1950).

[a] Discovered in 1971 by Loren Westphal. Complete specimen owned by UCLA. Stony with interstitial iron; density = 3.25 g/cc.

at one per 1 million square kilometers per year (Wasson 1985). Since the land surface of Iowa is about 150,000 square kilometers, the average rate of fall in Iowa should be about one fall in every six to eight years. Yet only four falls have been recorded by eyewitnesses since 1847, although many more probably have occurred. That more meteorites have not been found is because impact sites are hard to locate, even when fireballs are seen, and 95 percent of all falls are stones, which look like terrestrial basaltic igneous rocks. These meteorites might easily be confused with basaltic glacial erratics, which are abundant throughout the state.

The most common minerals occurring in iron meteorites are kamacite (Fe) and taenite (Fe-Ni). The minerals in stony meteorites are primarily magnesium and iron-silicates. Small amounts of mineral sulfides, like troilite (FeS), have also been reported.

2. Collecting Iowa's Minerals

When one thinks about states in which minerals are collected, Iowa is certainly not near the top of the list. In fact, it probably does not even make the list. Iowa is much better known for its fossils (e.g., the famed crinoids from Le Grand and the brachiopods and corals from the Rockford Brick and Tile site). Two reasons account for Iowa's low ranking as a state in which to collect minerals. First, the abundance and variety of minerals in the state are limited. In the world's famous gem and mineral localities (e.g., South Asia, Brazil, Russia, South Africa, and California), minerals are contained primarily in igneous and metamorphic rocks. Rocks of these types do not occur at the surface in Iowa, except as pebbles in stream gravels and scattered boulders in glacial deposits. The lone exceptions are the bedrock exposures of Precambrian Sioux Quartzite in Lyon County. Second, with the exception of quartz, minerals formed in Iowa are not very durable and therefore do not survive the harsh physical and chemical weathering conditions at the surface. Despite these limitations, museum-quality mineral specimens have been, and continue to be, extracted from limestone and shale bedrock exposures at several localities, particularly in the eastern part of the state. In this chapter, I will discuss how and where minerals are exposed for collecting, factors in mineral durability, how to access mineral-collecting sites, and some tips on mineral collection and preservation.

How and Where Minerals Are Exposed

Iowa's minerals are exposed by natural processes and human activities. In nature, minerals and the host rocks that enclose them are exposed primarily by the erosive action of Iowa's major rivers, such as the Mississippi, Missouri, Des Moines, and Cedar. Erosion by running water is controlled by river discharge, by climate-controlled rises and falls in the level of the ocean into which all Iowa rivers ultimately empty, and by tectonic uplift

2.1. Devonian fossil gorge. Coralville Dam, Johnson County, August 10, 1993. Note uneroded soil and glacial deposits at upper right.

and subsidence of the land surface. Rates of erosion are generally low, thus exposure of new material for collecting occurs slowly. Short bursts of more rapid erosion occur when rivers flow at peak discharge, for instance, during periods of rapid snow melt or unusually high rainfall. Spectacular examples of rapid erosion can be seen at the outflow areas below the Coralville and Saylorville dams, where uncontrolled discharge resulting from the prodigious rainfalls of 1993 scoured up to 25 meters or more of soil and bedrock from downstream river valleys (fig. 2.1). At the Coralville Dam near Iowa City, scouring exhumed a wealth of marine fossils and groundwater-sculpted limestone surfaces, which have attracted tens of thousands of visitors since the floodwaters receded. At Saylorville, north of Des Moines, fossils and minor pyritic mineralization are exposed in the gorge cut through the sandstone, shale, and limestone bedrock below the dam.

Minerals occurring in the limestone, shale, and sandstone outcrops along Iowa's rivers and creeks generally have been exposed for a very long time (tens to hundreds of thousands of years). For this reason, long-term preservation of minerals in natural outcrops requires that the minerals be durable. Crystalline carbonate rocks and well-cemented sandstones resist weathering, mass wasting, and fluvial erosion, and they stand in vertical to nearly vertical exposures, which are not easily covered by soil and vegetation. Among the many examples in Iowa are Silurian dolostones at Palisades Kepler State Park along the Cedar River east of Cedar Rapids,

2.2. Ordovician carbonate rocks near Dubuque.

Pennsylvanian sandstones and Mississippian limestones along the Des Moines River valley, Devonian dolostones at Backbone State Park near Strawberry Point, Ordovician carbonate rocks and sandstones along the Mississippi River from McGregor to Bellevue, and Devonian, Mississippian, and Pennsylvanian rocks along the Mississippi River from Davenport to Keokuk (fig. 2.2). Though these rocks appear to stand up well, the minerals they contain are only weakly resistant to the physical and chemical forces of weathering, and they are soon damaged or destroyed. In just a few years of exposure to Iowa's harsh climate, calcite dulls and cracks, and pyrite oxidizes (rusts). Minerals contained in weak shales, siltstones, and poorly cemented sandstones generally slough off along more gentle slopes and are soon covered with soil and vegetation.

Fortunately for collectors, Iowa's minerals are also exposed during highway, railroad, and building construction and by surface and underground mining. Many of the best localities for studying Iowa geology are along highway road cuts, for example, the Wyoming Hill cut along State Highway 22 near Muscatine (Pennsylvanian sandstones and shales), the entrance to Bellevue State Park in Jackson County (Ordovician shales and limestones) (fig. 2.3), the State Highway 9 cut from Churchtown to Lansing (Cambrian and Ordovician carbonate rocks), and the U.S. Highway 52 cut near Guttenberg (Ordovician carbonate rocks and sandstones). Road cuts also expose less durable shales and siltstones, which may be collected while the cut is still fresh. At Iron Hill near Waukon, the site of a formerly commercial iron-mining operation, rebuilding of U.S. Highway 9 in the early 1980s exposed masses of soft Cretaceous goethite and hematite adjacent to the old surface workings. Today, the exposure is overgrown with vegetation, and specimens are difficult to find.

Active quarries and mines provide the best opportunities for mineral collecting in the state. Not only is new material continually available for collecting, but exposures are fresh and minerals have not had time to experience the ravages of chemical and physical weathering. In quarries and open pits, fresh rock is blasted, removed, and crushed soon after exposure (fig. 2.4). Many superb mineral specimens have been reduced to fragments in rock-crushing plants in Iowa and elsewhere before they could be found and preserved, which makes those that have been spared the crusher's fate all the more prized. In underground mines, most of which operate on the room and pillar method, roof-supporting pillars remain as potential collecting areas long after the room material is removed (fig. 2.5). Pillars also provide many more rock surfaces for collecting than do the walls of quarries. During Iowa's mining history, zinc, lead, coal, gypsum, and limestone have been mined underground, and these types of mines have yielded collectible mineral specimens.

2.3. Limestone and shale outcrop. Bellevue State Park, Jackson County. This outcrop exposes rocks of the Maquoketa Shale (Ordovician).

2.4. Typical limestone quarry in eastern Iowa.

2.5. Drilling equipment at a working face of the Linwood Mine, Buffalo, Scott County. *Linwood Mining and Minerals Corporation.*

Mineral Durability

Most minerals in Iowa are products of events that occurred many millions of years ago. Many of the finest minerals collected to date formed in sedimentary diagenetic or hydrothermal environments at relatively shallow depths in the Earth, probably during the late Paleozoic or early Mesozoic eras (270–180 million years ago). This estimate is based on established ages of similar-appearing mineral deposits in Wisconsin, Illinois, and Missouri (e.g., see Brannon et al. 1992). To be preserved for collectors and researchers, minerals must survive postformational chemical attack. The attack comes chiefly from the action of surface and underground waters containing active chemical agents such as oxygen, carbon dioxide, and naturally occurring acids. Sulfide minerals form under chemically reducing (low oxygen activity) conditions. When exposed to oxygen-rich atmospheres for prolonged periods of time, these minerals oxidize to form new minerals, like limonite and smithsonite, which generally are less attractive to collectors than are their sulfide parents. Minerals with a high solubility in water, such as gypsum, anhydrite, and the iron sulfates, dissolve when in prolonged contact with water. Effects of dissolution range from etching and pitting of crystal faces to complete destruction of the original mineral (fig. 2.6). Minerals that are relatively immune to attack by water and oxygen alone become quite unstable when the water contains dissolved acid. Natural acids may form through the chemical union of water and atmospheric or soil-generated carbon dioxide (carbonic acid [H_2CO_3]) or from the oxidation of fine-grained pyrite (sulfuric acid [H_2SO_4]). Pyrite is abundant in organic-rich shales and coal beds. Exposure of these rocks to the atmosphere results in rapid decomposition of the pyrite and the generation of what is referred to as acid rock or acid mine drainage, a serious environmental problem in surface coal mining areas because the acid-laden waters and the dissolved chemicals they contain are toxic to plant and animal life. The carbonates, such as calcite and dolomite, are particularly susceptible to attack by acid-containing waters.

2.6. Calcite, chemically etched along cleavage and twin planes. Specimen is 6.5 centimers long. *Anderson Museum, Cornell College.*

Minerals must also survive physical attack from abrasion in stream-beds, expansion/contraction effects of freezing water and thawing ice, and shock pressures caused by blasting in mines, quarries, and road cuts. Minerals that possess low hardness and/or well-developed cleavage are most susceptible to physical damage. Such minerals bruise easily and must be collected with great care. Calcite, the most commonly occurring mineral in carbonate rocks, is soft and has excellent cleavage and thus is easily damaged. Next I will consider the durability of the more common minerals found in Iowa.

Anglesite ($PbSO_4$) is slightly soluble in acids. It is soft and brittle and therefore is damaged easily.

Barite ($BaSO_4$) is very slightly soluble in water and acid, hence it resists chemical attack. It dissolves slowly in warm concentrated sulfuric acid (H_2SO_4), leaving no common decomposition products. Barite is soft and has well-developed cleavage. It also commonly occurs in Iowa in bladed to waferlike crystals. These in particular are extremely fragile.

Calcite ($CaCO_3$), aragonite ($CaCO_3$), dolomite ($CaMg[CO_3]_2$) and other carbonate minerals have relatively low solubility in water but high solubility in acids. Dolomite is less soluble than calcite and aragonite. All are soft and have well-developed cleavage. Large calcite crystals are often damaged by blasting, and they are easily bruised during collecting. Exposure at the surface causes crystal surfaces to become dull and cracked.

Cerussite ($PbCO_3$) is soluble in acids. Crystal faces are quickly dulled when in contact with HCl. It is soft, brittle, and heat sensitive.

Fluorite ($CaF_2\cdot$) is slightly soluble in water and acid, thus it resists etching. On exposure to air, brown varieties may become clouded to nearly opaque, as observed in specimens from the Waterloo South Mine and Peske Quarry, but x-ray powder diffraction analysis does not reveal the presence of any alteration product (Garvin, unpublished data). Fluorite is quite soft and has well-developed cleavage; therefore, it must be handled carefully.

Galena (PbS), a sulfide of lead, has a very low solubility in water but is soluble in acids, especially HCl and HNO_3. It decomposes under oxidizing conditions to anglesite [$PbSO_4$] or cerussite [$PbCO_3$]. Once a thin oxide coating has formed, the remaining galena is generally protected from further decomposition. Galena is soft, has well-developed cleavage, and crumbles easily.

Gypsum ($CaSO_4\cdot2H_2O$) and anhydrite ($CaSO_4$) have relatively high solubility in water, and they are soluble in acids. These calcium sulfate minerals are soft and have well-developed cleavage. Complete, etch-free crystals are rare.

Halotrichite, rozenite, melanterite, and other hydrated iron sulfates form as decomposition products of iron sulfide minerals. In Iowa they are found typically as coatings and crusts on coal seams and pyrite-rich black shale. They are all highly soluble in water and can only survive where protected from surface runoff. Melanterite decomposes readily to rozenite unless kept in a humidified atmosphere.

Fine-grained, earthy varieties of hematite (Fe_2O_3) are slightly soluble in acids, especially HCl. They are also soft and disaggregate easily. Coarsely crystalline varieties resist chemical and physical attack.

Fine-grained varieties of limonite—a mixture of iron oxides, especially goethite [$HFeO_2$] and jarosite [$KFe_3(SO_4)_2(OH)_6$]—are slightly soluble in acids. They are also soft and disaggregate easily. Coarsely crystalline varieties, including pseudomorphs after pyrite and marcasite, are quite durable.

Millerite (NiS) is slightly soluble in water and acid. It oxidizes to a green nickel oxide that may coat crystal surfaces and stain adjacent calcite. Because of its capillary crystal habit, millerite is extremely fragile.

Pyrite (FeS_2), marcasite (FeS_2), and chalcopyrite ($CuFeS_2$) are relatively stable in water, provided that the free oxygen content is low. In oxygen-rich water or atmosphere they oxidize to form limonite. Under suitable conditions, the chemical replacement is pseudomorphic. The replacement typically occurs from crystal surface to core, and large crystals may have pseudomorphic rims but unaltered cores of original iron sulfide. Under relatively dry conditions of weathering, fine-grained pyrite may decompose to one or more of the hydrated iron sulfates (e.g., halotrichite, rozenite, and melanterite). In fresh exposures of shales containing pyrite nodules, erosion of soft shale may expose and concentrate nodules for easy collecting. Pyrite and marcasite are hard and lack cleavage; therefore, they resist bruising, except the fine-grained varieties. Chalcopyrite is soft and bruises easily.

Quartz, chalcedony, and chert (SiO_2) are insoluble in water and acids, except in hydrofluoric acid (HF). Chemically, quartz is the most stable of all the minerals that form in Iowa. It is also hard and lacks cleavage; therefore, it resists bruising and is otherwise physically very durable. The microcrystalline variety—chert—may alter through partial dissolution to a chalky substance, which often forms rinds around chert nodules. These rinds are soft and are easily damaged. X-ray powder diffraction analysis of this chalky substance reveals that it is microcrystalline quartz.

Smithsonite ($ZnCO_3$), like the other carbonates, is readily soluble in acids. Crystals are soft and cleave easily.

Sphalerite ([Zn,Fe]S) is insoluble in water but soluble in strong acid. It decomposes under oxidizing conditions to smithsonite [$ZnCO_3$] or hemimorphite [$Zn_4(Si_2O_7)(OH)_2 \cdot H_2O$]. Because of its solubility in acid, even when unoxidized, it is rarely found in crystals that are not etched. Sphalerite is soft and has well-developed cleavage; therefore, it bruises easily.

Durability information for these and other common minerals in Iowa is summarized in table 2.1. Relative chemical durability measures how well

Table 2.1. Chemical and Physical Durability of Common Iowa Minerals

Mineral	Chemical Formula	Decomposition Products	Resistance to Oxidation	Solubility in Water[a]	Solubility in Acids[b]	Hardness[c]	Cleavage	Relative Chemical Durability	Relative Physical Durability
Anglesite	$PbSO_4$		high	0.0042 (25)	sl H_2SO_4	3–3.5	3 pinacoidal	high	low
Anhydrite	$CaSO_4$		high	0.209 (30)	s	3–3.5	3 pinacoidal	low	low
Aragonite	$CaCO_3$		high	0.0015 (25)	s	3.5–4	1 pinacoidal	low	low
Barite	$BaSO_4$		high	0.0002 (25)	sl H_2SO_4	3–3.5	3 pinacoidal	high	low
Calcite	$CaCO_3$		high	0.0014 (25)	s	3	3 rhombohedral	low	low
Cerussite	$PbCO_3$		high	0.0001 (20)	s	3–3.5	3 pinacoidal	low	low
Chalcopyrite	$CuFeS_2$	limonite, malachite	low			3.5–4	none	moderate	low
Dolomite	$CaMg(CO_3)_3$		high	0.032 (18)	s	3.5–4	3 rhombohedral	low	low
Iron sulfates			high	generally high	sl to s	<4	no data	low	low
Fluorite	CaF_2		high	0.0016 (18)	sl	4	4 octahedral	high	low
Galena	PbS	anglesite, cerussite	moderate	0.0124 (20)	s	2.5	3 cubic	moderate	low
Goethite	$FeO(OH)$		high	i	s HCl	5–5.5[d]	1 pinacoidal	high	microcrystals: low macrocrystals: high

Mineral	Formula	Alteration products		Solubility[a]	[b]	Hardness[c]	Cleavage		
Gypsum	$CaSO_4 \cdot 2H_2O$		high	0.241 (25)	s	2	3 pinacoidal	low	low
Hematite	Fe_2O_3	limonite	high	i	sl–s	5–6[d]	none	high	microcrystals: low macrocrystals: high
Marcasite	FeS_2	limonite, Fe-sulfates	low	0.0005 (25)	i	6–6.5	none	microcrystals: low macrocrystals: high	high
Millerite	NiS	bunsenite	low	0.0004 (18)	s	3–3.5	3 prismatic	moderate	low
Pyrite	FeS_2	limonite, Fe-sulfates	low	0.0005 (25)	i	6–6.5	none	microcrystals: low macrocrystals: high	high
Quartz	SiO_2		high		i except HF	7	none	high	high
Sphalerite	ZnS	smithsonite, hemimorphite	low	0.00006 (18)	vs	3.5–4	6 dodecahedral	moderate	low

[a] Values in grams per cm³. Numbers in parentheses indicate temperatures in degrees Celsius at which solubility determinations were made.

[b] s = soluble, vs = very soluble, sl = slightly soluble, i = insoluble.

[c] Hardness numbers are in reference to standard Moh's scale.

[d] Earthy varieties appear to be much softer.

Source: Solubility data from *Handbook of Chemistry and Physics*, 56th ed., Chemical Rubber Company, Cleveland.

the mineral will survive attack by the natural elements of chemical weathering. Relative physical durability measures how easily the mineral may be damaged during erosion, quarrying, mining, road construction processes, and collecting. For further information on mineral durability see Sinkankas (1970, 1972).

Accessing Collecting Localities

Collectors should understand and respect the laws and observe the courtesies pertaining to ownership of the land on which mineral-collecting sites are located. On some public lands, such as national parks and monuments and state parks and preserves, natural areas are preserved for the "public good," and rock, mineral, and fossil collecting are prohibited. Violators are subject to prosecution. Collecting along public roadways is generally permitted, with the exception of interstate highways. Avid mineral collectors are often oblivious to everything but the minerals they are seeking, including vehicular traffic. Be alert when collecting along roadways.

Collecting on private land requires permission of the property owner. Most private landowners are very cooperative with mineral collectors, provided that permission is asked and that property is respected. Once a collecting area has been located, the identity of the property owner can be determined from a county plat book, copies of which are available at county office buildings and at many banks and real-estate agencies. Once permission to collect is granted, collectors must be respectful of fences, gates, farm equipment, and other personal property. Nothing will sour a property owner to future mineral collecting more than a damaged fence or a gate left open, allowing valuable livestock to escape. After the collecting is complete, it is a good policy to express appreciation for the opportunity, either by stopping at the owner's house or by sending a letter. Some owners, especially those in the geode areas of southeast Iowa, charge a fee for collecting on their lands (Horick 1974).

As stated earlier, in general the best mineral collecting is done in active quarries and mines. Quarry and mine operators own or lease the land. Because of the hazards and liabilities that attend underground mining processes, mine operators are very reluctant to permit collecting underground. In matters of safety, the mining industry is regulated by the Mining Safety and Health Administration (MSHA), an agency of the federal government. Access to underground mines requires instruction in hazard identification and avoidance, even for visitors. Most mine operators do not have sufficient staff to provide training on a regular basis, nor the self-rescuing equipment that must be carried by every person going underground. MSHA inspects mines regularly and levies stiff fines for violations of its safety regulations, including untrained personnel underground. Some mining companies will arrange for supervised group tours of their underground operations on an occasional basis, but generally for observation rather than collecting.

Gaining access to quarries is somewhat easier; however, quarries are also hazardous. MSHA regulates quarry safety, but hazards training for surface personnel is less complicated than that for those going underground. Still, for liability reasons some quarry owners will not grant permission to individual collectors. Organized mineral-collecting groups, such as the Cedar Valley Rock and Mineral Society and the Black Hawk Gem and Mineral Society, have been able to gain access to some quarries. One may also gain access to collecting areas by participating in organized geologic field trips, such as those conducted twice a year by the Geological Society of Iowa. When permission is granted, visitors are generally required to wear hard hats, hard-toed shoes, and safety glasses, and they must sign a waiver to absolve the company from liability in case they suffer an accident while on the property. If there is to be any hope of mineral collecting at quarries in the future, it is imperative that collectors respect the dangers present in quarries. In addition, they must not interfere with vehicular traffic and other quarrying operations, and they must respect company property, including machinery, buildings, and drainage and pumping systems. Documented cases record fences torn down, gate locks sawed through or broken, machinery and tools damaged or stolen, windows smashed, and trash left behind. As an example of why mine and quarry operators become frustrated with collectors, several years ago at a quarry in eastern Iowa the foreman showed up on Monday morning to find that many of the large rip-rap boulders placed along the quarry rim to ensure the safety of vehicular traffic had been pounded to gravel by a group of eager collectors the previous Saturday.

When in quarries, collectors should observe the following recommendations.

Wear hard hats, safety glasses, and, if possible, steel-toed safety boots.

Respect quarry walls. Although most walls are scaled (loose rocks are removed with long pry bars), many quarry walls contain unstable material. Unstable walls are especially dangerous during the spring and fall when freezing and thawing causes rocks to be plucked out of the walls (fig. 2.7). When a wall looks bad, stay away from it. If a half-ton rock falls on your head, a hard hat will not save you.

Beware of quarry rims. Unstable walls are also dangerous when one is standing on a rim. Because of vibrations due to blasting, the rock can be severely cracked as much as 5 meters back from the edge. Resist the urge to survey the quarry from the rim.

Beware of water hazards. In active quarries, the floors are engineered so that water drains toward the lowest level (sump), from which water is pumped to the surface beyond the quarry rim (fig. 2.8). Sump water can

2.7. High wall in limestone quarry. Note numerous cracks and other indications of instability.

2.8. Draining sump in limestone quarry.

2.9. Portable rock-crushing plant.

be deep—5 meters or more. The sides of the sump are in many cases nearly vertical. In abandoned quarries, the water may be as much as 7 meters deep. Drownings of swimmers in old quarry ponds are all too frequent. Save your swimming for the pool.

Watch out for trips, slips, and falls. These are the most common accidents in quarries. Collectors become so involved in looking for specimens that they fail to watch their own feet. Quarry floors are generally rough and irregular. Rock piles on quarry floors often contain boulders that dislodge when stepped on carelessly. Watch where you step.

Beware of rock-crushing and -hauling equipment. Pedestrians do not have the right-of-way in quarries. If you are permitted to be in a quarry during times of rock removal, stay away from crushers and haulage roads (fig. 2.9).

Do not handle any wires that may be lying among the loose rocks. They were probably used to connect explosives to a detonator. Occasionally a charge does not explode, and the exposed wire might still be connected to it.

Tips on Mineral Collection and Preservation

Experienced collectors generally know where to look for good minerals and what tools to use to extract them from the enclosing rock. Local rock and mineral–collecting groups can provide a wealth of knowledge of this kind. For the novice, the following information may be helpful.

Locating Collecting Sites

For collectors who are unfamiliar with collecting localities, maps can be very helpful. To be really useful, a map must contain sufficient detail to

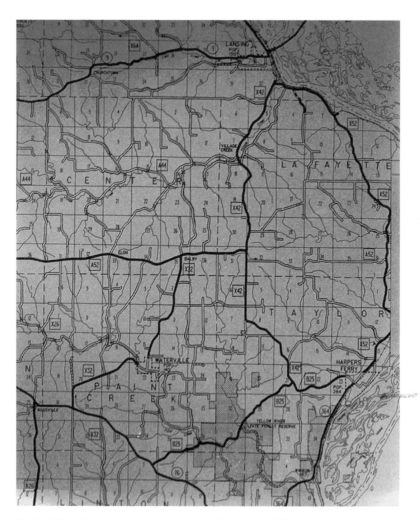

2.10. Highway and transportation map of a portion of Allamakee County. The numbered squares are a mile across. *Iowa Department of Transportation, 1991.*

identify all hard-surface and gravel roads, stream drainages, and cultural objects, such as buildings, parks, and quarries.

County highway maps issued by the Iowa Department of Transportation show all roadways and major streams. They may show quarries and gravel pits. Since the maps are not printed in color and since they emphasize roads, streams and other features may be difficult to distinguish (fig. 2.10).

Topographic maps are prepared and issued by the U.S. Department of the Interior's Geological Survey (USGS). Like county highway maps, they show roadways, stream drainages, and cultural objects. In addition, by means of contour lines, which connect points of equal elevation on the land surface, they illustrate the nature of the topographic surface—such as locations of hills and valleys and steep versus gentle slopes. The different features of the topographic map are shown in different colors. Water is blue; vegetation is green; contour lines are brown; cultural objects are red, black, or fuschia. The standard topographic map covers a much smaller land area than does a county highway map; therefore, it can present information in greater detail. Learning to read a topographic map thoroughly requires a little time and instruction. Most elementary-level textbooks in physical geology have a section at the back that explains the fundamentals of map reading. Topographic maps of any area in Iowa can be obtained at the Geological Survey Bureau, Iowa Department of Natural Resources in Iowa City (fig. 2.11).

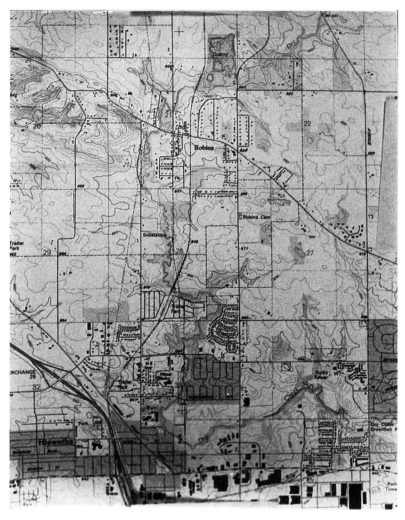

2.11. Topographic map of an area north of Cedar Rapids, including Robins Quarry (north, center). The numbered squares are a mile across. *U.S. Geological Survey Cedar Rapids North, Iowa Quadrangle, 7.5 Minute Series, 1994.*

2.12. Mineralized fracture. The white calcite indicates that other minerals may be present.

Geologic maps possess all the information contained on a topographic map. In addition, by means of color overprinting, the kinds of rocks that appear at or near the land surface are shown. These maps are useful in identifying the locations of rock units that are favorable hosts for specific minerals, for example, Warsaw Shale, which contains geodes. A geologic map of the state of Iowa is shown in figure 1.3 and plate 15.

County plat books, which show the locations of parcels of land and identify the current owner(s), are very helpful when one is trying to obtain permission to enter private property for the purpose of collecting minerals. They are available for observation at county planning and development offices, at local banks, and at real-estate offices.

Recognizing Favorable Host Rock

Collectible minerals in Iowa commonly line small cavities or fill fractures in carbonate, siltstone, or shale. Calcite is the most abundant mineral in the vast majority of these localities and is generally colorless, white, or pale yellow. Calcite-bearing mineralization stands out as localized white areas in a matrix of gray to brown limestone or dolostone (fig. 2.12). Finding calcite may lead to other less common minerals, such as marcasite, sphalerite, barite, and fluorite. Pyrite frequently is recognized by its oxidation product, limonite, which stains the adjacent rocks an intense rust color.

Field Equipment

Successful mineral collecting requires proper field gear. The experience will be more pleasant and satisfying if you dress properly. The excitement of a mineral-collecting adventure certainly is dulled by skinned knees, sliced fingers, sunburned skin, or rain-drenched clothing. Finding the best mineral specimens in Iowa generally requires climbing among slabs and blocks of rock. Pants made of heavy denim or comparable fabrics and sturdy shoes or boots can take the punishment of rough, sharp-edged rocks in addition to protecting tender skin. Gloves help protect hands from cutting by sharp crystals and from blows meant for the cold chisel. Horsehide or cowhide gloves are the most durable, cloth gloves the least. A hard hat and safety goggles or glasses are a must; in fact, they are required for entrance into most quarries. Also, carry a first-aid kit. Skinned knuckles (and knees, if you insist on wearing shorts) are common in quarries and at natural rock outcrops. Sufficient layers of insulating clothing for cold weather, sunblock, and a canteen of water will help make the collecting trip more enjoyable.

A waterproof jacket, preferably with a hood, is another essential, because a good time to collect minerals in Iowa is during or immediately following a rainstorm. Water washes dust and other foreign material from rock and mineral surfaces, making the minerals easier to see. This suggestion is particularly applicable when one is collecting from gravel pits. Water brightens rock surfaces and increases contrasts among rock and mineral types, making identifying good specimens easier. Some of the best collecting, both from bedrock and gravel, is done during a light drizzle.

The most important tools of the mineral collector's trade are those used to separate desired mineral specimens from their enclosing host rocks. They include sledges, rock hammers, chisels, gads, and prying instruments. Detailed descriptions of these tools and how to use them can be found in Sinkankas (1970). Remember that each tool is designed to perform a specific function. Do not use a pry bar or a rock hammer as a chisel. Trying to drive the pointed end of a rock hammer into a crevice by striking a blow with another rock hammer can damage both hammers, because both are made of tempered steel. It is also dangerous because splinters of steel can fly off during impact.

A good mineral collector is careful to document all specimens collected. A small notebook and pencil will fit conveniently into the collecting bag. Good documentation should include the name of the collecting site, its map location, the stratigraphic unit in which the minerals occur, and the mode of their occurrence (e.g., cavity lining, fracture fill, breccia cement, replacement of host rock). A waterproof marker can be used to place an identification number on an inconspicuous place on the specimen.

Collecting and Preserving the Minerals

Geodes generally are collected loose from streambeds. The hard chalcedony rind protects the more fragile crystal linings from bumping and abrasion during stream transport. The most attractive geodes are hollow and are lined with quartz and other minerals. If the hollow space in the geode is large, the geode will feel unusually light. Hollow geodes sometimes rattle because of the presence of loose crystals inside. Remember that impacts of loose crystals may damage fragile crystal linings. Avoid breaking the geode in the field. Hollow geodes generally shatter on impact. Resist the urge to find out what is inside until the geode can be opened properly with a diamond saw or geode cracker.

Selecting material for removal from a crystal-lined cavity in bedrock requires a trained eye. Best-quality specimens consist of unbroken, bruise-free crystals. As previously noted, calcite, barite, fluorite, and galena commonly form in cavities. All of these minerals are soft and have excellent cleavage; therefore, they are easily damaged. Some collectors have the mistaken idea that bigger is better. Far more important than size are freedom from damage and the aesthetic characteristics of the specimen, such as arrangement of shapes and colors. Removal of minerals from cavities takes patience and time—up to several hours for a top-quality specimen. In active quarries fragile crystals are frequently damaged by shock from blasting and rock falls. Undamaged crystals are more often the exception. Years ago in the Conklin Quarry I discovered in a loose limestone block a cavity lined with large (up to 6 centimeters long), pale yellow calcite crystals. The crystals were mostly damage-free because, by good fortune, the cavity was completely filled with soft clay, which had absorbed the shock of blasting. When prying fragile crystals loose from cavity walls, first pack the cavity with newspaper, burlap, "bubble" paper, or other cushioning material to lessen damage due to vibration or sudden dislodging.

Once the mineral specimens are removed from the host rock, they must be packed for transport to the home or laboratory. Fragile minerals should be wrapped individually in protective material. Newspaper, heavy paper toweling, foam plastic sheets, and "bubble" paper all work well. If the collecting site is a long distance from the vehicle, wrapped specimens can be placed in cloth or paper sacks and carefully inserted into a pack. If the vehicle is close, wrapped specimens can be loaded in boxes for easy transport. Cardboard flats ("beer flats") work well because minerals can be packed one layer deep and stored conveniently under a vehicle seat. Very fragile materials (like millerite) that should not be in contact with anything but air should be girdled with packing material and the tops left uncovered. Small specimens store well in Styrofoam egg cartons.

Identifying minerals in the field or laboratory can, at times, be challenging, even for experienced collectors. Pyrite and marcasite are often confused. Pyrite has been misidentified as chalcopyrite, acicular marcasite as millerite, and brown rhombohedral calcite as fluorite. Mineral-collecting guides are helpful, but they often suffer from two defects: (1) they show ideal samples with well-formed, undamaged crystals, which may not look much like the garden-variety specimen you have found; and (2) because of imperfections in color printing, color fidelity is often poor. When in doubt about the identity of a mineral, seek help from more experienced collectors and professional mineralogists. Appendix C lists mineral-collecting organizations in Iowa and colleges and universities that employ professional mineralogists. A 10-power hand lens is the constant companion of all serious students of rocks and minerals. Good-quality lenses can be obtained from any supplier of outdoor and laboratory geology equipment.

Once safely in the home or laboratory, mineral specimens should be cleaned and labeled. How much cleaning should be done is largely a matter of personal preference. It should be remembered that *any* cleaning removes natural material. Some collectors prefer to preserve their minerals in their natural condition, that is, as they obtained them at the source. For many minerals the color and luster of the mineral can be improved greatly by washing and, in some cases, by treatment with chemicals.

Simple washing will remove most dirt and plant debris and some oxide films. Washing, or at least rinsing, should be done with distilled water, because tap water (hard and soft) will leave a mineral residue that can dull lustrous crystal faces. Removing foreign material from pits and cracks requires spraying the specimen with water under pressure or brushing and picking. Most minerals may be wet- or dry-brushed. For hard minerals like quartz and coarsely crystalline pyrite, a hard-bristle brush (like a toothbrush) works well. Lifting out material beyond the reach of the brush can be done with a sharpened dental probing tool or a sewing needle mounted in a handle. For soft minerals like calcite and barite, a soft-bristle toothbrush or a camelhair artist's brush will do. Sinkankas (1970) advocates the use of a slivered bamboo skewer, because the bamboo is both soft and strong and is capable of penetrating tiny recesses. Delicate minerals like millerite and acicular marcasite can be cleaned effectively in an ultrasonic cleaner. Sinkankas (1970) describes in detail the operation of this device. Fragile minerals like halotrichite and other iron sulfates, due to their softness and high solubility in water, should probably be left untouched.

Some coatings cannot be removed by brushing. These require treatment with chemicals that will dissolve the unwanted coating while at the same time leaving the mineral unaffected. Rust stains on quartz and calcite can be removed effectively by soaking the specimen for several hours in a solution of oxalic acid and water. Oxalic acid powder can be purchased from a paint store or pharmacy. The powder and solution are poisonous when ingested, and the solution can irritate the skin. Always work with safety goggles and rubber gloves. After removal from the acid bath, the specimen should be bathed in successive distilled water baths in order to remove unsightly yellow stains (Sinkankas 1972). Calcite is somewhat soluble in oxalic acid, and prolonged exposure will dull and eventually pit calcite crystal faces. Some iron oxide coatings, such as the iridescent films on pyrite and marcasite and the golden stains on quartz in some geodes, may increase the attractiveness of the specimen and therefore should not be removed. Unwanted calcite coatings on quartz and other minerals can be removed by carefully applying a dilute solution (5 to 10 percent) of hydrochloric acid (HCl). Hydrochloric acid gives off corrosive vapors; therefore, it should only be used outdoors or with a well-ventilated fume hood. Less powerful acids that will also, if more slowly, dissolve calcite include acetic acid and citric acid. When using any corrosive chemical, always wear goggles and rubber gloves. Before applying any acid, always test an inconspicuous area of the mineral to be certain that the acid does not react with the mineral. Detailed information on the use of chemical reagents and their effects on individual minerals is found in Sinkankas (1970, 1972).

Mineral specimens, like photographs, should be labeled to identify their contents and sources. It is frustrating to encounter a mineral specimen or, worse, a mineral collection with no information about where the minerals were found. Each mineral specimen should contain an identifying number that is keyed to a paper label that identifies the mineral or minerals present, the source, and the collector's name. A good way to create a permanent label is to put a small spot of white modeling paint in an inconspicuous place on the specimen (or on an enclosing container if the specimen is too small). When the paint is thoroughly dry, write the identifying number with permanent black ink. Then cover the ink with a spot of clear fingernail polish, which will protect the number when the specimen is handled. If desired, the information from each labeled mineral specimen can be entered into a computer database for easy retrieval.

Some minerals are difficult to preserve once they are removed from their natural setting. Atmospheric and other conditions in a mineral display case are generally very different from those under which the mineral formed. Some minerals, like certain varieties of fluorite, are sensitive to sunlight or artificial light. Their colors will fade or change unless they are stored in the dark. Other minerals are sensitive to changes of atmospheric moisture. The iron sulfate melanterite, which forms underground in a moist, oxygen-abundant atmosphere, will decompose in as little as a few hours in the typically drier atmosphere at the surface. Fine-grained (or sooty) pyrite is very unstable in humid air. This mineral is frequently found in pyritic nodules occurring in coal-bearing strata in southern Iowa and in paleokarst-associated black shales in eastern Iowa. The accumulation of white powder in cracks and on the surface of the specimen is a signal that decomposition of sooty pyrite is occurring. The chemical decomposition of this material produces sulfuric acid, the corrosive vapors of which may cause the disintegration of cardboard storage containers and the degradation of nearby minerals, especially those that are reactive to acid. I learned from sad experience that corrosive vapors from the disintegration of pyrite in my own mineral display case caused irreversible blackening of nearby malachite [$Cu_2CO_3(OH)_2$] and aurichalcite

$[(Zn,Cu)_5(CO_3)_2(OH)_3]$ and etching and rust-colored staining of contiguous calcite. Attempts to stabilize the pyrite—for example, by soaking it in bleach (which is supposed to kill decomposition-causing bacteria)—generally have proven unsuccessful. The only guarantee for preservation of fine-grained pyrite is storage in a container that is free of atmospheric moisture.

3. Occurrences of Iowa's Minerals

This chapter describes the major mineral occurrences in Iowa. Most of these mineral deposits are exposed in active or abandoned quarries and mines. For more information regarding host rock stratigraphy, refer to the generalized geologic column (table 1.1). Minerals resulting from the oxidation of preexisting species are not included in the descriptions unless they constitute more than thin coatings on the host minerals. Commonly occurring oxidation minerals are limonite, hematite, malachite, smithsonite, anglesite, and cerussite. Ultraviolet fluorescence for selected minerals was determined using a long-wave (365 nm) source. Additional crystallographic data, including illustrations of commonly occurring crystal forms, are given in Appendix A. A guide to mineral identification can be found in Appendix B.

Agate and Chert Localities

Agate, the name given to banded, microfibrous quartz, can be found in a wide variety of localities in Iowa. The different types of agate are distinguished primarily by their color, by the nature of their banding, and by whether they occur in bedrock or river gravel sources. I will describe the more well known agates.

Lake Superior agate *occurrence*: in gravel deposits along the Mississippi River and along major rivers and creeks coursing across the eastern half of Iowa; the agates were transported by glacial ice and running water from sources near Lake Superior in northern Wisconsin; *crystallography*: microfibrous quartz (chalcedony); concentric ("fortification") banding; agates up to 15 centimeters long have been reported, though the great majority range from 1 to 4 centimeters long; *color*: typically alternating red-to-brown and white-to-gray bands (fig. 3.1).

3.1. Lake Superior agates. Cedar River near Ivanoe Bridge, Linn County. Central front pebble is 3 centimeters across. *Anderson Museum, Cornell College.*

Notes: Lake Superior agates are prized by collectors and lapidarists. While these agates might be encountered along any stream point bar, good places to find them are in the piles of coarse gravel that accumulate at commercial sand and gravel plants. Gravel pits on the Iowa side of the Mississippi River near Bellevue and Muscatine are well-known sources. Of course, permission is required before entering these properties.

References: Menzel and Pratt 1969; Sinkankas 1970; Horick 1974.

Coldwater agate *occurrence*: as nodular replacements of or cavity fillings in carbonate rock; unlike Lake Superior agate, so-called coldwater agate originated in Iowa; it has been collected from several localities in Benton (Brandon-Urbana area), Black Hawk, Bremer, Delaware, Linn, and Keo-kuk counties; *crystallography*: microfibrous quartz (chalcedony), exhibiting concentric banding; pea- to football-sized; *color*: blue-gray to white but may exhibit red, orange, or black coloration due to staining by iron or manganese oxides (fig. 3.2).

References: Menzel and Pratt 1965; Horick 1974.

Rice agate *occurrence*: actually chert, this variety of microcrystalline quartz occurs as nodules in the Plattsmouth Limestone (Pennsylvanian),

3.2. Coldwater agate. Near Urbana, Benton County. Specimen is 12 centimeters across. *Anderson Museum, Cornell College.*

3.3. Rice agate. Montgomery County. Specimen is 5.5 centimeters across. *Ray Anderson collection.*

which result from replacement of fusilinid-bearing limestone (fusilinids are wheat or rice grain–shaped protozoans); *crystallography*: microgranular quartz (chert); *color*: fossils are light gray to light blue; the "rice" matrix is dark gray to buff (fig. 3.3).

Notes: Good sites for collecting rice agate are quarries near Stennet and Grant in Montgomery County. The compactness of the chert varies; in less compact varieties the "rice" may pluck out during grinding and polishing.

———

References: Menzel and Pratt 1963; Horick 1974.

Buffalo Quarry

Location: Scott County, at Buffalo, just west of the Linwood Mine; Sec. 17, T77N, R2E (fig. 3.4).

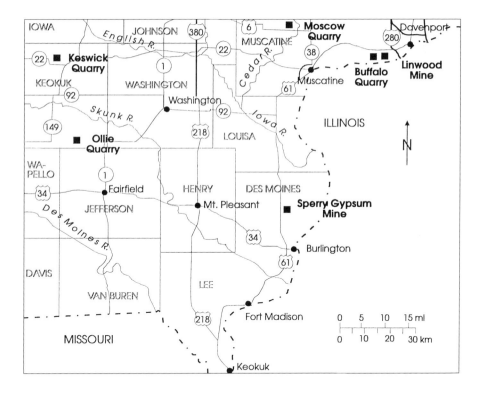

3.4. Major mineral occurrences in southeast Iowa.

Host Rock Stratigraphy: Middle Devonian; floor—Spring Grove Member, Pinicon Ridge Formation; rim—Coralville Formation; paleokarst fillings of early Pennsylvanian sediments.

Modes of Mineral Occurrence: cavity linings, fracture linings and fillings, breccia cements, nodules and disseminations in organic-rich paleokarst-filling mudstones and replacements of woody plant material contained therein.

Minerals

barite *abundance*: uncommon; *occurrence*: fracture lining, breccia cement; *crystallography*: bladed, up to 5 centimeters long; *color*: colorless, milky white, pale yellow; *UV fluorescence*: nonfluorescent.

calcite *abundance*: abundant; *occurrence*: cavity linings, fracture fillings, breccia cement; *crystallography*: acute scalenohedra with obtuse rhombo-hedral phantoms, multiple acute scalenohedra in parallel intergrowths, sparry where present in fracture fillings; *color*: colorless to pale yellow, multiple variety is nearly opaque white; *UV fluorescence*: colorless type—dull pink; white type—white base and bright pink top (fig. 3.5 and pl. 1).

dolomite *abundance*: common; *occurrence*: cavity and fracture linings; *crystallography*: obtuse rhombohedral with curved faces, individual crystals up to 3 millimeters across; *color*: milky to creamy white.

3.5. Calcite on pyrite. Buffalo Quarry, Scott County. Specimen is 10.5 centimeters long. *Anderson Museum, Cornell College.*

marcasite *abundance*: common; *occurrence*: fracture linings, breccia cement, inclusions in calcite; *crystallography*: bladed, wedge-shaped, complexly twinned, individual crystals up to 0.5 centimeter long; *color*: metallic greenish gray, iridescent surfaces due to oxidation.

pyrite *abundance*: common; *occurrence*: fracture linings, breccia cement, nodular masses and disseminations in paleokarst-filling mudstones and sandstone, replacement of woody plant material, inclusions in calcite; *crystallography*: cubo-octahedral, octahedral, microcrystalline, rarely stalactitic, individual crystals up to 0.5 centimeter across; *color*: brass yellow, widely exhibiting iridescent tarnish due to oxidation.

sphalerite *abundance*: uncommon; *occurrence*: disseminations in paleokarst-filling mudstone, replacement of host rock; *crystallography*: poorly formed tetrahedral crystals that are highly etched, sparry masses, individual crystals up to 3 centimeters across; *color*: dark brown, yellow brown on thin edges.

Notes: Buffalo Quarry is very near the Linwood Mine (which is off limits to collectors), and it exposes many of the same stratigraphic units as does Linwood. Excellent specimens of calcite and marcasite have been found here. Pyrite, though crystals are small, is attractive because of its iridescence.

———

Reference: Garvin, unpublished data.

Cedar Rapids Quarry

Location: Linn County, west of Palisades Kepler State Park and south of U.S. Highway 30; Sec. 15, T82N, R6W (fig. 3.6).

Host Rock Stratigraphy: Silurian; floor—Scotch Grove Formation, rim—Gower Formation; minor paleokarst-filling sediments of Pennsylvanian age.

Modes of Mineral Occurrence: dissolution-enlarged high-angle fracture linings and fillings; breccia cements; vug linings and fillings.

Minerals

calcite *abundance*: abundant; *occurrence*: vug fillings (type A)—lower part of quarry; fracture linings and fillings (type B)—upper part of quarry; *crystallography*: type A calcite crystals—acute scalenohedra more or less modified by rhombohedra, with crystals ranging from a few millimeters to a few centimeters in length; type B—sparry or acute scalenohedral, with crystals ranging up to 10 centimeters in length; rare rhombohedral twins;

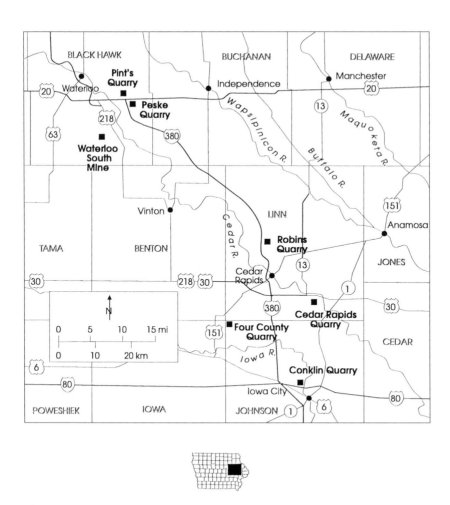

3.6. Major mineral occurrences in east-central Iowa.

color: type A calcite—gray to amber to chocolate brown; type B—white to amber; *UV fluorescence:* type A brown varieties—dull to bright red; gray varieties—nonfluorescent; type B—dull to bright pink (fig. 3.7).

chert *abundance:* abundant; *occurrence:* nodular and lensoid replacements of dolostone, chert is especially abundant in the lower unit (Scotch Grove); crystallography: Scotch Grove—nodules and lenses, the edges of which may appear ragged due to replacement by dolomite; Gower—nodules up to 5 centimeters in diameter that are smooth-surfaced, with some nearly perfectly spherical; *color:* white, gray, black—the black is due to the presence of finely disseminated iron sulfide minerals (fig. 3.8).

marcasite *abundance:* common; *occurrence:* fracture linings in crusts up to 6 centimeters thick, disseminations in and replacements of host

3.7. Vug lined with marcasite (dark rim) and calcite. Cedar Rapids Quarry, Linn County. Specimen is 9 centimeters across. *Anderson Museum, Cornell College.*

3.8. Chert nodule in dolostone matrix. Cedar Rapids Quarry, Linn County. Nodule is 2.5 centimeters in diameter. *Anderson Museum, Cornell College.*

dolostone; *crystallography*: well-formed single blades up to 0.5 centimeter long, complex twins and intergrowths; *color*: pale brass yellow on freshly broken surfaces, rust colored on oxidized surfaces.

pyrite *abundance*: common; *occurrence*: associated with marcasite in crusts, inclusions in type B calcite; *crystallography*: generally microcrystalline exhibiting cube modified by octahedron and pyritohedron; *color*: brass yellow on fresh surfaces, rust colored on oxidized surfaces.

quartz *abundance*: uncommon; *occurrence*: linings of small vugs in lower part of quarry; *crystallography*: drusy; *color*: white.

sphalerite *abundance*: common; *occurrence*: exclusively in fracture linings in upper part of quarry, crusts up to 2 centimeters thick, associated with type B calcite; *crystallography*: generally sparry but locally as etched malformed tetrahedral crystals; *color*: yellow brown to brown.

Notes: Large calcite crystals are locally abundant. The crystal surfaces are variably weathered, and crystal clusters are particularly fragile when wet. Good marcasite crystals also occur but are uncommon. Occasional good brown calcite crystals can be found in the lower part of the quarry. The chert nodules from the Gower Formation are spectacular because of their high degree of sphericity. They are easily removed from the host dolostone.

———

References: Garvin 1984a; Garvin et al. 1987; Garvin and Ludvigson 1993; Garvin, unpublished data.

Coal-Associated Minerals of South-Central Iowa

Locations: coal-mining areas of southern Iowa, particularly near Lovilia in Monroe County, Knoxville in Marion County, Oskaloosa in Mahaska County, and Eddyville in Wapello County (fig. 3.9).

Host Rock Stratigraphy: shale, limestone, and coal units of the Pennsylvanian Des Moines Series; mineral-bearing limestones occur as nodular and lensoid masses that are locally referred to as coal balls.

Modes of Mineral Occurrence: shale—nodules, disseminations, and fracture fillings; coal—disseminations and fracture fillings (cleats); limestone—fracture fillings in coal balls.

Minerals

barite *abundance*: uncommon; *occurrence*: fracture linings in coal balls, perched on calcite; *crystallography*: tabular, very small rosettes; *color*: small

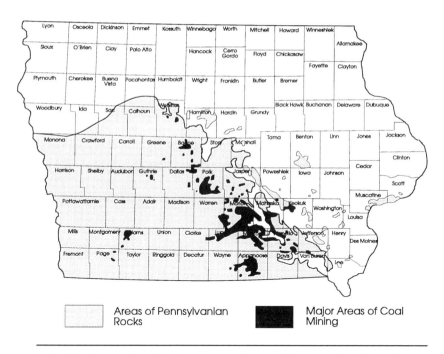

3.9. Distribution of coal-bearing strata and areas of former coal mining in Iowa.

Areas of Pennsylvanian Rocks

Major Areas of Coal Mining

rosettes—salmon pink; tabular crystals—colorless, white, locally yellow cores with salmon pink rims; color variations suggest two generations; *UV-fluorescence*: nonfluorescent.

calcite *abundance*: abundant; *occurrence*: fracture linings in coal balls and linings of dessication cracks in septarian nodules; *crystallography*: early—obtuse and acute rhombohedra up to 1 centimeter long, some rhombs are attenuated or curved; late—single and multiple acute scalenohedra up to several centimeters long; *color*: early—colorless to dark brown, locally with iridescent red, green, and lavender coatings; late—colorless to white; *UV fluorescence*: early—bright to dull creamy white, exhibits prominent banding; late—nonfluorescent.

goethite *abundance*: uncommon; *occurrence*: perched on calcite in coal balls; *crystallography*: long-prismatic to acicular; *color*: black (pl. 2).

gypsum *abundance*: abundant; *occurrence*: single crystals and aggregates in gray and black shales; *crystallography*: prismatic crystals exhibiting familiar beveled diamond shape, locally contain phantom crystals highlighted by carbon inclusions, elongated single crystals and swallow-tail twins up to 25 centimeters long.

malachite *abundance*: rare; *occurrence*: in septarian nodules; *crystallography*: short, wiry filaments; *color*: green.

marcasite *abundance*: common; *occurrence*: inclusions in coal ball calcite, considerably less common than pyrite; *crystallography*: long prismatic, bladed; *color*: greenish gray.

pyrite *abundance*: abundant; *occurrence*: crystals, nodular masses, disseminations in black shale; fracture fillings and lenses in and replacements of coal; coal breccia cement; inclusions in coal ball–associated calcite; coatings on limestone; *crystallography*: shale—octahedron, modified by cube, up to several centimeters across; locally octahedra are stacked in parallel growths radiating outward as much as 15 centimeters in several directions from a center; coal—massive fine-grained; coal balls—scattered microcrystals in and coatings on calcite.

quartz *abundance*: common; *occurrence*: fracture linings in coal balls; *crystallography*: doubly terminated crystals up to 3 centimeters long; resemble hexagonal dipyramid with minor hexagonal prism development; *color*: colorless to smoky black (pl. 2).

siderite *abundance*: common; *occurrence*: fracture linings in coal balls, impure beds (?); *crystallography*: flat obtuse rhombohedra up to 1 centimeter across; *color*: light to dark brown.

sphalerite *abundance*: uncommon; *occurrence*: cleat fillings in coal; *crystallography*: sparry; *color*: dark brown.

Notes: The best mineral crystals are associated with coal balls and organic-rich shales, and these are found frequently in coal strip mines. The coal balls are easily recognized as large masses of dark gray rock cast aside during the mining process. A representative collection of coal ball minerals is displayed at the University of Iowa's Department of Geology in Iowa City.

References: Keyes 1893; Gricius 1964; McCormick and Bailey 1973; Horick 1974.

Conklin Quarry

Location: Johnson County, at Coralville, north of Interstate 80, on west side of Iowa River; Secs. 32 and 33, T80N, R6W (fig. 3.6).

Host Rock Stratigraphy: Middle Devonian; floor—Spring Grove Member, Pinicon Ridge Formation; rim—Coralville Formation; paleokarst fillings of early Pennsylvanian Caseyville Formation.

Modes of Mineral Occurrence: limestone—linings and fillings of dissolution cavities, high-angle fracture fillings, breccia cements, disseminations in

3.10. Cluster of calcite crystals. Conklin Quarry, Johnson County. Specimen is 11 centimeters across. *Anderson Museum, Cornell College.*

limestone and associated organic-rich shale partings; paleokarst (mudstones and sandstones)—disseminations in organic-rich mudstone, sandstone cement, coatings on contained limestone clasts.

Minerals

barite *abundance*: common; *occurrence*: cavity linings and fracture fillings; *crystallography*: tabular to bladed, blades may reach 2 centimeters or more in length; *color*: colorless to pale yellow; *UV fluorescence*: nonfluorescent.

calcite *abundance*: abundant; *occurrence*: cavity linings, fracture fillings, breccia cements; *crystallography*: several generations—early forms rhombohedral, commonly with brown coatings, later forms acute scalenohedral; basal and rhombohedral twins; crystals may reach several centimeters in length; *color*: colorless to pale yellow; *UV fluorescence*: early rhombohedral—creamy white; late scalenohedral (white)—dull to bright pink; late scalenohedral (colorless)—nonfluorescent (fig. 3.10 and pl. 4).

chalcopyrite *abundance*: uncommon; *occurrence*: scattered on and included in calcite; *crystallography*: modified pseudotetrahedral, generally too small for forms to be recognized without magnification; *color*: metallic yellow on freshly broken surfaces, reddish tarnish (fig. 3.11).

dolomite *abundance*: rare; *occurrence*: lining small vugs; *crystallography*: curved rhombohedral ("saddles"), up to 4 millimeters across; *color*: creamy white.

3.11. Chalcopyrite crystals perched on calcite. Conklin Quarry, Johnson County. Crystals are 3 millimeters across. *Anderson Museum, Cornell College.*

marcasite *abundance*: common, less abundant than pyrite; *occurrence*: encrustations on and inclusions in calcite, scattered on Davenport limestone, disseminated in paleokarst mudstone; *crystallography*: typically bladed and wedge-shaped twins and intergrowths, rare rosettes; *color*: metallic greenish yellow on freshly broken surfaces (pl. 3).

millerite *abundance*: common; *occurrence*: perched on and included in calcite; *crystallography*: crystals are needlelike (acicular), commonly in brushlike groups that project from calcite crystal surfaces, individual needles may exceed 4 centimeters in length; *color*: metallic bronze on fresh surfaces, oxidation produces green coatings and stains in surrounding calcite (fig. 3.12).

pyrite: *abundance*: abundant; *occurrence*: encrustations on and inclusions in calcite, linings of cave walls, nodular masses in karst-filling mudstone, disseminations in mudstone and limestone, sandstone cement, replacement of woody Pennsylvanian plant material; *crystallography*: cubic and cubo-octahedral, also microcrystalline, cubes in mudstone may reach 3 centimeters across; *color*: brass yellow, fine-grained varieties oxidize quickly on exposure to air to form rust-colored limonite coatings on quarry walls (pl. 3).

quartz *abundance*: common; *occurrence*: in ellipsoidal masses up to 20 centimeters in length in lower Davenport Formation; *crystallography*: microcrystalline, interiors commonly cellular in structure; *color*: colorless, pores may contain scattered calcite and mineral sulfides.

sphalerite *abundance*: common; *occurrence*: cavity linings on calcite, fracture fillings, disseminations in shale partings, in pyrite nodules in mudstone; *crystallography*: typically sparry crystal habit, occasionally tetrahedral forms that are strongly etched, sparry masses may exceed 4 centimeters across; *color*: dark brown to yellow brown, very lustrous on freshly broken surfaces.

Notes: The Conklin Quarry is one of the better-known localities in the state for collecting minerals. It is famous for its millerite, but museum-quality specimens of calcite and pyrite have also been found. The quality of specimens has declined in recent years as quarrying moves southward.

———

References: Horick 1974; Garvin 1984a; Garvin et al. 1987, Garvin and Ludvigson 1988, 1993; Kutz and Spry 1989.

Dubuque Area Lead-Zinc Mines

Location: The lead-zinc mines in Iowa are part of a much larger mining district, commonly known as the Upper Mississippi Valley Zinc-Lead District, the center of which is in southwestern Wisconsin. Mining in Iowa was concentrated along the western margin of the Mississippi River, principally in and adjacent to Dubuque (fig. 3.13). In fact, abandoned mines underlie portions of the city. Some mining was also done in eastern Clayton County northwest of Guttenberg and between the Turkey River and

3.12. Millerite. Conklin Quarry, Johnson County. Specimen is 7 centimeters across. *Anderson Museum, Cornell College.*

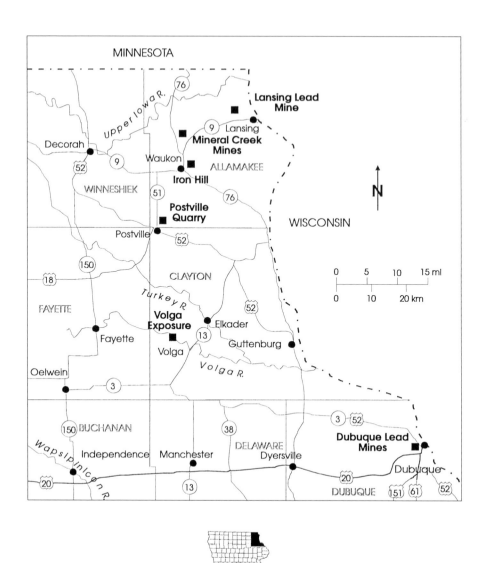

3.13. Major mineral occurrences in northeast Iowa.

North Buena Vista. In Dubuque County, in addition to the mining at Dubuque, mines were located near Sherrill and in the Sageville-Durango area. A few small mines operated at Tete des Mortes in extreme northeastern Jackson County.

Host Rock Stratigraphy: chiefly within the Ordovician Galena Dolomite, with minor occurrences in the underlying Decorah, Platteville, and Prairie du Chien formations.

Modes of Mineral Occurrence: solution-enlarged vertical fractures (gash veins or crevices), low-angle transverse and bedding plane fractures (pitches and flats), breccia cements.

Minerals

anglesite *abundance*: common; *occurrence*: coatings on weathered galena, unaltered galena cores may be present; *crystallography*: microcrystalline; *color*: varying shades of gray.

aragonite *abundance*: uncommon; *occurrence*: on roofs and walls of caves; *crystallography*: fibrous ("satin spar"), radiating, helictitic masses; *color*: white, colorless.

barite *abundance*: rare; *occurrence*: reported from a water well and a stream deposit near Dubuque, also in early alluvial deposits; barite, though uncommon, is widespread in other parts of the Upper Mississippi Valley Zinc-Lead District.

calcite *abundance*: abundant; *occurrence*: fracture linings and fillings, breccia cements, cave linings; *crystallography*: macrocrystalline—acute scalenohedra modified by one or more obtuse rhombohedra, individual crystals 15 centimeters or more in length; microcrystalline—acicular, fibrous, stalactitic, banded masses (travertine or "onyx"); *color*: macrocrystalline—colorless to milky white; microcrystalline—colorless to white.

cerussite *abundance*: common; *occurrence*: coatings on weathered galena; *crystallography*: microcrystalline; *color*: white to blue-gray.

dolomite *abundance*: uncommon; *occurrence*: lining of small cavities in host carbonate rock; *crystallography*: rhombohedral crystals 1 or 2 millimeters across; *color*: tan, gray.

galena *abundance*: abundant; *occurrence*: fracture linings and fillings, cavity linings, breccia cements; *crystallography*: cubic ("cog lead"), cubo-octahedral, individual crystals up to 18 centimeters across weighing 27 kilograms, fracture fillings are anhedral masses ("sheet lead"); *color*: crystals rarely lustrous but rather exhibit a dull gray, corroded appearance; commonly coated with light gray to white anglesite or cerussite or with limonite (fig. 3.14).

gypsum *abundance*: uncommon; *occurrence*: crevice fillings, locally on cave floors; *crystallography*: sparry, fibrous, acicular; acicular crystals up to 15 centimeters in length reported on a cave floor near Dubuque; *color*: colorless, white.

3.14. Galena crystals ("cog lead"). Dubuque lead mines, Dubuque County. Specimen is 12 centimeters across. *Anderson Museum, Cornell College.*

marcasite *abundance*: abundant; *occurrence*: similar to pyrite; *crystallography*: masses of bladed and wedge-shaped crystals, frequently in subhedral aggregates, crystals up to 1 centimeter or so in length; *color*: pale brass yellow but, like pyrite, commonly altered to limonite.

pyrite *abundance*: abundant; *occurrence*: linings of fracture walls, breccia cement; *crystallography*: commonly as subhedral aggregates; where well crystallized, crystals are octahedral or cubo-octahedral, crystals up to 2 centimeters across; *color*: brass yellow but frequently in various stages of alteration to rust brown to brown limonite.

smithsonite *abundance*: abundant; *occurrence*: cavity linings, fracture linings and fillings, breccia cements, replacements of fossils and other host carbonate rock, product of near-surface decomposition of sphalerite; *crystallography*: spongy, cellular masses of microcrystals, commonly referred to as "dry bone" or "sponge bone," locally banded and colloform; *color*: tan, brown, gray.

sphalerite *abundance*: common; *occurrence*: similar to smithsonite, commonly occurs as unaltered cores in masses of smithsonite, encountered in deeper levels of mining; *crystallography*: compact layers and sparry masses, locally banded; *color*: dark brown to nearly black.

Notes: The Dubuque Area Lead-Zinc Mines are no longer accessible to collectors, and most dumps have been reclaimed. Representative minerals are on display at the University of Wisconsin-Platteville.

References: Leonard 1894, 1896; Calvin and Bain 1899; Shipton 1916; Heyl et al. 1959; Horick 1974; Ludvigson and Dockal 1984.

Fort Dodge Gypsum Deposits

Location: Webster County, near Fort Dodge (fig 1.5).

Host Rock Stratigraphy: gypsum beds are the host rock; they are determined to be Jurassic in age, based on gymnosperm pollen analysis; they rest unconformably on Mississippian and Pennsylvanian units and are overlain by fine-grained sandstones and red and green shales of Jurassic or Cretaceous age.

Mode of Geological Occurrence: in horizontal beds, which are commonly laminated.

Minerals

celestite *abundance*: uncommon; *occurrence*: masses in layers several centimeters thick between shales; *crystallography*: columnar, sparry, fibrous; *color*: light blue. This mineral was reported by Keyes (1893) but not in any of the more recent publications (Bard 1982; Dasenbrock 1984; Hayes 1986).

gypsum *abundance*: abundant; *occurrence*: in horizontal beds exhibiting fine-scale laminations, also as horizontal veins; *crystallography*: microcrystalline in beds, fibrous in veins; in associated organic-rich shales tabular selenite, exhibiting swallow-tail twins, has been reported; *color*: colorless, white with gray laminations, reddish near top of deposit (fig. 3.15).

miscellaneous minerals *abundance*: rare; the deposit consists of more than 90 percent gypsum; minor minerals include calcite, dolomite, quartz, clay minerals, and orthoclase.

Notes: Gypsum crystals are generally microscopic; laminated masses are attractive, especially when slabbed and polished.

References: Keyes 1893; Spencer 1895; Wilder 1902; Horick 1974; Bard 1982; Dasenbrock 1984; Hayes 1986; Cody et al. 1996.

3.15. Laminated gypsum. Fort Dodge, Webster County. Specimen is 16 centimeters high. *Anderson Museum, Cornell College.*

Four County Quarry

Location: northwest corner of Johnson County, southwest of Cedar Rapids Municipal Airport; Sec. 4, T81N, R8W (fig. 3.6).

Host Rock Stratigraphy: Middle Devonian; floor—Solon Member, Little Cedar Formation; rim—Coralville Formation; paleokarst fillings of Pennsylvanian age.

Modes of Mineral Occurrence: fracture linings and fillings and cave wall linings associated with large paleokarst fill; nodules, disseminations, and sandstone cements, replacement of Pennsylvanian woody plant material; coatings on and replacement of contained clasts and cave walls; scattered small vugs, fracture fillings, and disseminations in limestone elsewhere in the quarry; many similarities with the Conklin Quarry.

Minerals

barite *abundance*: rare; *occurrence*: vug linings; *crystallography*: rosettes of tabular crystals; *color*: white; *UV fluorescence*: nonfluorescent.

calcite *abundance*: abundant; *occurrence*: fracture fillings and vug linings; *crystallography*: sparry, acute scalenohedral, vug-associated crystals commonly etched and may reach 3 centimeters in length; *color*: colorless to white; *UV fluorescence*: bright orange-pink (fig. 3.16).

chalcopyrite *abundance*: rare; *occurrence*: vug linings, scattered on and in calcite crystals; *crystallography*: microscopic pseudotetrahedal crystals; *color*: metallic yellow, reddish yellow on tarnished surfaces.

dolomite *abundance*: rare; *occurrence*: vug linings; *crystallography*: rhombohedral crystals with curved faces; *color*: tan.

marcasite *abundance*: common; *occurrence*: mainly as coatings on paleo-karst-contained limestone clasts; *crystallography*: bladed, wedge-shaped, and complexly twinned; maximum crystal length a few millimeters; *color*: pale brass yellow on freshly broken surface, rust colored where oxidized (fig. 3.17).

millerite *abundance*: rare; *occurrence*: vug linings, inclusions in calcite crystals; *crystallography*: acicular crystals; *color*: metallic bronze.

pyrite *abundance*: abundant; *occurrence*: fracture linings, coatings on paleokarst-contained limestone clasts, nodules in paleokarst-filling mudstone and cement in sandstone, nodules up to 6 centimeters in length, replacement of carbonized wood fragments, replacement of limestone adjacent to paleokarst-filling mudstones; *crystallography*: cubic and cubo-octahedral, commonly occurring in scattered clusters on limestone, also microcrystalline, maximum crystal size 1.5 centimeters; *color*: brass yellow on freshly broken surface, rust colored where oxidized.

Notes: The best specimens reported from the Four County Quarry are pyrite nodules, which reach several centimeters in length and are very lustrous. These are found in soft black mudstone and are easily removed. Calcite crystals are generally quite weathered, in part due to acids produced from exposure of the paleokarst deposits to moisture and air. This

3.16. Calcite crystals. Four County Quarry, Johnson County. Specimen is 10 centimeters across. *Anderson Museum, Cornell College.*

3.17. Marcasite rosettes on limestone. Four County Quarry, Johnson County. Specimen is 15 centimeters across. *Anderson Museum, Cornell College.*

quarry is also famous for its trilobites, and museum-quality specimens have been removed from the limestones.

References: Garvin et al. 1987; Garvin and Ludvigson 1988, 1993; Kutz and Spry 1990.

Geode-Associated Minerals of Southeast Iowa

Locations: Geode-bearing strata are exposed along the Mississippi River and tributary drainages in southeastern Iowa and adjacent northern Missouri and western Illinois. Geodes have been collected from localities in the following Iowa counties: *Des Moines*: along Long Creek about 3.2 kilometers northeast of Augusta (Sec. 18, T69N, R3W); *Henry*: along Mud Creek east of Lowell to the creek's confluence with the Skunk River (Secs. 22 and 27, T70N, R5W); "Baltimore Hills," adjacent to Geode State Park about 8 kilometers south of New London; *Lee*: along Soap Creek at Keokuk; along a small stream 3.2 kilometers northwest of Denmark (Sec. 19, T69N, R4W); in a quarry along Lamalees Creek north of Sandusky (NW 1/4, Sec. 36, T66N, R5W); *Van Buren*: along Indian Creek across the river from Farmington (Sec. 3, T67–68N, R8W); along Bear Creek 0.8 kilometer south of Vernon (Sec. 11, T68N, R9W); along Copperas Creek about 3.2 kilometers northwest of Bentonsport (Sec. 26, T69N, R9W). Geodes have also been collected near Niota, Nauvoo, Warsaw, and Hamilton, Illinois, and near Wayland and Fox City, Missouri.

Host Rock Stratigraphy: Mississippian; Warsaw and Keokuk formations, especially the shale-rich horizons.

Modes of Mineral Occurrence: linings and fillings of spheroidal, flattened ellipsoidal, and irregular-shaped cavities; locally with multiple cavities; the cavities are believed to result from replacement and dissolution of calcareous concretions that had previously formed in shale-rich horizons in dolostone; cementation of crushed geode fragments; replacement of dolomitic host rock; geodes range in size from less than a centimeter to more than a meter in diameter; spheroidal cavities tend to have hollow interiors, while flattened cavities are generally filled.

Minerals

aragonite *abundance*: uncommon; *occurrence*: white crusts on earlier minerals; *crystallography*: stellate groups of acicular crystals, anhedral crusts; *color*: white; *UV fluorescence*: greenish yellow, rarely pale rose.

barite *abundance*: common; *occurrence*: projecting from earlier minerals; *crystallography*: thin to thick tabular crystals, rosettes of tabular crystals, crystals up to 8 centimeters across; *color*: sky blue to white, rarely pale green, blue variety exhibits color zoning.

calcite *abundance*: common; *occurrence*: scattered crystals and linings on quartz, interior bands between outer chalcedony and quartz; *crystallography*: modified obtuse rhombohedra up to several centimeters across, acute scalenohedra up to 10 centimeters long, local phantoms highlighted by iron sulfide inclusions, local organic films; *color*: colorless to milky white, pink, dark brown, locally rust colored due to staining by iron oxide; *UV fluorescence*: early anhedral linings—greenish yellow to white; later euhedra—nonfluorescent.

chalcopyrite *abundance*: rare; *occurrence*: perched on and included in calcite; *crystallography*: pseudotetrahedral crystals ranging up to 2 millimeters across; *color*: metallic yellow, with coatings of green malachite or black tenorite.

dolomite *abundance*: common; *occurrence*: scattered on quartz, on chalcedonic rinds (this dolomite may be unreplaced host rock); *crystallography*: aggregates of rhombohedra commonly exhibiting curved crystal faces, crystals up to a few millimeters across; *color*: tan, light pink to rose red, white, locally dark brown (may be ankerite).

goethite *abundance*: uncommon; *occurrence*: perched on quartz crystals; *crystallography*: acicular crystals up to 7 millimeters long, subparallel aggregates; *color*: black.

gypsum *abundance*: common; *occurrence*: scattered crystals and encrusting films on quartz or chalcedony; *crystallography*: elongated prismatic crystals up to 4 centimeters long, stellate aggregates, microscopic films; *color*: colorless to white.

hematite *abundance*: uncommon; *occurrence*: perched on quartz or calcite crystals, also in interior zones in quartz; *crystallography*: earthy coatings, rarely obtuse rhombohedral crystals up to 1 millimeter across; *color*: coatings—red to pink; rhombohedra—blood red, with brilliant metallic luster.

kaolinite *abundance*: uncommon; *occurrence*: interstitial fillings of quartz and other cavity-lining minerals; *crystallography*: flocculent powder; *color*: white.

marcasite *abundance*: common; *occurrence*: inclusions in calcite, scattered on earlier minerals; *crystallography*: long prismatic and bladed microcrystals, acicular crystals up to 25 millimeters long; *color*: metallic white to greenish gray, commonly pseudomorphically replaced by limonite.

millerite *abundance*: uncommon; *occurrence*: inclusions in and projections from calcite, rarely perched on quartz; *crystallography*: acicular habit, individual whiskers and tufted or matted aggregates up to 6 centimeters long; *color*: metallic bronze; some capillary marcasite has been misidentified as millerite.

pyrite *abundance*: common; *occurrence*: scattered on all earlier minerals, inclusions in calcite, disseminations in host rock shale; *crystallography*: cube, locally modified by octahedron and pyritohedron, individual crystals up to 10 millimeters across, capillary pyrite has also been reported; *color*: brass yellow, commonly with dark brown to rust brown coatings of limonite due to oxidation, locally iridescent.

quartz *abundance*: abundant; type I, microcrystalline (chalcedony)—*occurrence*: rinds on outer surface of geode (this is virtually a universal characteristic of Iowa geodes and gives the geode resistance against physical and chemical weathering); rather than resulting from initial deposition on the walls of an open cavity, rinds are believed to result from replacement of outer margins of concretions; coatings on earlier macrocrystalline quartz; *crystallography*: microcrystalline, colloform, commonly banded, locally chalky due to alteration; thickness of rinds ranges from 1 millimeter to 2 centimeters or more; *color*: milky white to bluish gray, locally rust colored due to staining by iron oxide; type II, macrocrystalline (quartz)— *occurrence*: cavity linings and fillings, replacement of host rock; *crystallog-*

3.18. Quartz geode. Near Keokuk, Lee County. Specimen is 12 centimeters across. *Anderson Museum, Cornell College.*

raphy: hexagonal prism modified by obtuse rhombohedra; coarse to fine druse; massive, anhedral; individual crystals range from 1 millimeter to 3 centimeters or more in length; *color*: colorless, druse is locally rust colored due to staining by iron oxide (fig. 3.18).

sphalerite *abundance*: common; *occurrence*: perched on quartz or calcite, late cement in some crushed geodes; *crystallography*: anhedral masses and poorly formed tetrahedral crystals up to 8 centimeters across, cleavage masses up to 30 centimeters across, locally contains oriented inclusions of chalcopyrite; *color*: dark brown (fig. 3.19).

sulfur *abundance*: rare; *occurrence*: scattered on altered sphalerite or pyrite; *crystallography*: reported as disphenoids but probably dipyramids, up to 1 millimeter long; *color*: lustrous greenish yellow; may actually be a hydrated iron sulfate (szmolnokite).

Notes: Geodes are abundant along creek beds in Des Moines, Henry, Lee, and Van Buren counties, and they wash out of adjacent exposures during periods of high stream discharge. Most of the collecting areas are on private land. Some areas can be collected on a fee-for-collecting basis, others by permission of property owners. Geodes should be collected whole. Striking them with a rock hammer may shatter them, especially those that have hollow centers. They should be opened carefully with a rock saw or geode splitter.

3.19. Sphalerite (black) cementing crushed chalcedony geode rim. Near Keokuk, Lee County. Specimen is 9 centimeters across. *Anderson Museum, Cornell College.*

References: Savage 1902; Van Tuyl 1912, 1922; Tripp 1959; Hayes 1964; Sinnotte 1969; Horick 1974; Witzke 1987.

Iron Hill

Location: Allamakee County, about 4.8 kilometers northeast of Waukon; Sec. 17, T98N, R5W (fig 3.13).

Host Rock Stratigraphy: Cretaceous Windrow Formation.

Modes of Mineral Occurrence: fracture filling and replacement of chert-bearing limestone.

Minerals

goethite *abundance*: abundant; *occurrence*: fracture filling and replacement of limestone; *crystallography*: hard, nodular masses having a fibrous habit; coarse, cellular masses; soft, earthy masses; *color*: dark brown to yellow brown.

hematite *abundance*: abundant; *occurrence*: fracture filling and replacment of limestone; *crystallography*: earthy; *color*: red-brown.

miscellaneous minerals: in adjacent southeastern Minnesota, pyrite, illite, and siderite have been reported from similar-appearing exposures of ironstone; these minerals may also occur at Iron Hill, though they have not been reported.

Notes: Masses of ironstone up to a half meter across were exposed during the construction of a road cut on State Highway 9 adjacent to the old quarry in the early 1980s. The cut is largely overgrown with vegetation, and at present finding specimens is difficult.

References: Calvin 1895; Howell 1915; Bleifus 1972; Horick 1974.

Jackson County Goethite

Location: Jackson County, north of Andrew and east of Canton; Jones County, near Scotch Grove; large concentration in Sec. 19, T85N, R3E.

Host Rock Stratigraphy: all mineral material found as float; bedrock in the general area is Silurian dolostone (Hopkinton, Gower, and Scotch Grove formations).

Modes of Mineral Occurrence: minerals were not found in place, but the slabby, stalactitic nature of the material suggests that minerals formed on cave walls and ceilings; flat-backed slabs up to 30 centimeters long, with stalactitic protrusions, have been found.

Minerals

marcasite *abundance*: abundant; *occurrence*: cave and dissolution-enlarged fracture linings; *crystallography*: stalactitic masses of bladed crystals, blades up to 1 centimeter long, encrustations of twinned crystals up to 2 centimeters long; *color*: almost completely altered to goethite, tiny remnants of unaltered marcasite encountered during slabbing of material (Steve Barnett, personal communication), goethite is orange brown to dark brown.

pyrite *abundance*: abundant; *occurrence*: cave linings; *crystallography*: stalactitic masses of cubic and cubo-octahedral, crystals up to to 2 centimeters across; *color*: almost completely altered to goethite, tiny remnants of unaltered pyrite encountered during slabbing of material (Steve Barnett, personal communication), goethite is orange brown to dark brown (fig. 3.20).

Notes: Blocks and slabs of goethite pseudomorphs can be found along small stream drainages in the locations indicated.

References: Steven Barnett, personal communication; Garvin, unpublished data.

3.20. Stalactitic goethite pseudomorphs after pyrite. Near Andrew, Jackson County. Specimen is 12 centimeters across. *Anderson Museum, Cornell College.*

Keswick Quarry

Location: Keokuk County, about 1.6 kilometers west of Keswick; NW 1/4, Sec. 21, T77N, R12W (fig. 3.4).

Host Rock Stratigraphy: Mississippian; floor—Prospect Hill Formation; rim—Spergen Formation.

Modes of Mineral Occurrence: upper level—geodal fillings of cavities; lower level—small bedding plane–controlled cavities near chert layers and lenses.

Minerals

barite *abundance*: uncommon; *occurrence*: lower level—deposits on cavity-lining calcite; upper level—on friable red sandy material; *crystallography*: lower level—prismatic crystals up to 3 centimeters long; upper level—clusters of tabular crystals, individual crystals up to 7 centimeters across; *color*: prismatic crystals are transparent yellow, tabular crystals exhibit yellow cores with light to dark gray rims, dark color likely due to microscopic inclusions of iron sulfide; *UV fluorescence*: nonfluorescent.

calcite *abundance*: abundant; *occurrence*: cavity linings; *crystallography*: sparry, modified acute scalenohedral crystals up to 12 centimeters long, locally containing microinclusions of FeS_2, rarely as small stalactites (up to 4 centimeters long) that are completely enclosed in chalcedony; *color*: yellow to colorless; *UV fluorescence*: early sparry—lemon yellow; scalenohedral—pale pink (fig. 3.21).

chalcedony *abundance*: common; *occurrence*: cavity linings; *crystallography*: cryptocrystalline quartz, masses a half meter or more across; *color*: variable; red, brown, orange, pink, white are common; distributed fine-scale bands generally parallel to margins of cavity; this is the famous Keswick agate (pl. 5).

dolomite *abundance*: uncommon; *occurrence*: cavity linings, associated with sphalerite; *crystallography*: curved rhombohedra; *color*: white.

goethite *abundance*: uncommon; *occurrence*: cavity linings scattered on quartz; *crystallography*: microcrystalline, prismatic to acicular; *color*: black.

pyrite *abundance*: common; *occurrence*: scattered crystals and local clusters on calcite and quartz; *crystallography*: cubic; *color*: brass yellow, dark brown on weathered surfaces.

quartz *abundance*: abundant; *occurrence*: vug linings; *crystallography*: drusy; individual crystals 1 millimeter or so across; where associated with

3.21. Calcite crystals. Keswick Quarry, Keokuk County. Specimen is 10 centimeters across. *Anderson Museum, Cornell College.*

3.22. Drusy quartz dusted with tiny goethite crystals. Keswick Quarry, Keokuk County. Specimen is 9 centimeters across. *Anderson Museum, Cornell College.*

agate, crystals may be 2 centimeters or more long; *color*: colorless, white to smoky black (fig. 3.22).

sphalerite *abundance*: uncommon; *occurrence*: late deposits on cavity-lining calcite, chalcedony, or quartz; *crystallography*: sparry, subhedral clusters, complex twins up to 3 centimeters across with octahedral appearance; *color*: medium to dark brown.

Notes: The Keswick Quarry is best known for its banded agate. Red and brown are the most sought after colors. The drusy quartz from the lower level is very showy. Well-formed crystals of calcite, barite, and sphalerite are found occasionally, especially near the floor of the quarry.

––––––

References: Horick 1974; Garvin unpublished data.

Lansing Lead Mine

Location: Allamakee County, northwest of Lansing; Sec. 10, T99N, R4W (fig. 3.13).

Host Rock Stratigraphy: Ordovician; Hager City and Stockton Hill members, Oneota Formation.

Modes of Mineral Occurrence: dissolution-enlarged vertical fracture ("gash vein"); breccia cement.

Minerals

anglesite *abundance*: common; *occurrence*: coatings on galena; *crystallography*: microcrystalline; *color*: light gray but variably stained by limonite, results from oxidation of galena.

calcite *abundance*: uncommon; *occurrence*: small fracture and vug fillings; *crystallography*: sparry; *color*: colorless to white but variably stained by limonite; *UV fluorescence*: nonfluorescent.

cerussite *abundance*: common; *occurrence*: perched on galena in sheltered voids; *crystallography*: lustrous prismatic crystals up to 5 millimeters across; *color*: colorless to dark gray, results from oxidation of galena (fig. 3.23).

dolomite *abundance*: abundant; *occurrence*: replacement of limestone and linings in small vugs; *crystallography*: microcrystalline; *color*: gray-brown but variably stained by limonite.

galena *abundance*: abundant; *occurrence*: fracture fill; *crystallography*: generally massive and sparry, cubic in open vugs; *color*: bright metallic gray on freshly broken surfaces.

3.23. Cerussite crystals on weathered galena. Lansing Lead Mine, Allamakee County. Largest crystal is 4 millimeters across. *Anderson Museum, Cornell College.*

goethite *abundance*: common; *occurrence*: fracture fillings and coatings of breccia clasts; *crystallography*: microcrystalline; some goethite is a pseudomorphic replacement of cubic and octahedral pyrite and bladed marcasite, thus the crystal forms of the original iron sulfide minerals can be observed; *color*: rust brown to dark brown.

hemimorphite *abundance*: common; *occurrence*: interstitial fillings in galena; *crystallography*: microcrystalline; *color*: white but variably stained by limonite, results from oxidation of sphalerite.

quartz *abundance*: abundant; *occurrence*: breccia cement, vug linings; *crystallography*: drusy; *color*: colorless, may be stained red by hematite.

smithsonite *abundance*: common; *occurrence*: interstitial fillings in galena; *crystallography*: microcrystalline; *color*: white but variably stained by limonite, results from oxidation of sphalerite.

Notes: The Lansing Lead Mine was abandoned shortly after 1900. It can be located only by the careful observer because the terrain is heavily overgrown. The north-south trench and some of the waste piles are still visible. Occasional pieces of massive galena can still be found. The best specimens are those that contain cerussite crystals, some of which are large enough to be seen easily without magnification.

References: Calvin 1895; Leonard 1896; Heyl et al. 1959; Horick 1974; Ludvigson 1976; Garvin, unpublished data.

Linwood Mine

Location: Scott County, at Buffalo; Sec. 13, T77N, R2E (fig. 3.4).

Host Rock Stratigraphy: Middle Devonian; the mine exposes two stratigraphic units, the Cedar Rapids Member of the Otis Formation and the Davenport and Spring Grove members of the Pinicon Ridge Formation; paleokarst fillings of undetermined Pennsylvanian age.

Modes of Mineral Occurrence: linings of cavities; coatings of breakdown blocks; breccia cements; minor fracture fillings in limestone; replacements of limestone along cavity walls; disseminations in paleokarst-filling sandstones and mudstones; sandstone cement; replacements of woody plant material; growths in paleokarst-filling mudstone.

Minerals

barite *abundance*: common; *occurrence*: cavity linings, coating of breakdown blocks, growths in paleokarst-filling mudstone and altered limestone; *crystallography*: wide variety of crystal habits—tabular (single crystals and rosettes), plumose, drusy, prismatic; single barite prisms up to 12 centimeters in length; *color*: colorless, white, pale yellow to dark amber; *UV fluorescence*: multicrystalline barite—white to pale yellow; monocrystalline barite—nonfluorescent (pls. 6, 9).

calcite *abundance*: abundant; *occurrence*: cavity linings, coating of breakdown blocks, breccia cements, growth in paleokarst-filling mudstone; *crystallography*: five generations of calcite have been identified, the earliest is an acute scalenohedral druse, later generations consist of alternations of acute scalenohedral and obtuse scalenohedral and rhombohedral forms; phantoms are common and are highlighted by microscopic inclusions of iron sulfide minerals on interior growth surfaces; acute scalenohedra up to 20 centimeters in length and rhombohedra up to 12 centimeters across have been observed; rare rhombohedral, basal, and penetration twins; *color*: colorless to pale amber; *UV fluorescence*: earliest calcite—creamy yellow to white; later calcite—orange pink to dull violet (pls. 8, 10).

chalcopyrite *abundance*: rare; *occurrence*: perched on and included in calcite; *crystallography*: microtetrahedral crystals; *color*: metallic yellow on freshly broken surfaces, bronze on oxidized surfaces.

dolomite *abundance*: rare; *occurrence*: cavity linings; *crystallography*: rhombohedral crystals about 2 millimeters across; *color*: tan.

gypsum *abundance*: rare; *occurrence*: formed on fracture surfaces in dessicated organic-rich paleokarst-filling mudstone; *crystallography*: microprismatic; *color*: colorless.

marcasite *abundance*: abundant; *occurrence*: cavity linings, coating of breakdown blocks, breccia cements, growth in mudstone, disseminations in mudstone, common as microscopic inclusions in calcite; *crystallography*: scattered blades and wedges and rosettes, maximum crystal size 0.5 centimeter; *color*: pale metallic brass yellow on freshly broken surfaces, greenish gray on oxidized surface (pls. 7, 8, 10).

melanterite *abundance*: rare; *occurrence*: formed in organic-rich paleokarst-filling mudstone; *crystallography*: short prismatic, fibrous, microcrystalline; *color*: greenish blue, quickly alters to white powder on exposure to air; other hydrated iron sulfates (e.g., rozenite, siderotil, coquimbite, halotrichite) and epsomite likely occur in oxidized iron sulfide–rich mudstone; rozenite has been confirmed by x-ray powder diffraction analysis.

pyrite *abundance*: common; *occurrence*: cavity linings, microscopic inclusions in calcite, disseminations and nodular masses in paleokarst-filling

3.24. Calcite (white) perched on sphalerite (black). Linwood Mine, Scott County. Specimen is 8 centimeters across. *Anderson Museum, Cornell College.*

mudstone, replacement of limestone adjacent to mudstone-filled cavities, rare sandstone cement, replacement of carbonized woody plant material; *crystallography*: most commonly microcrystalline; where distinguishable, crystals are cubo-octahedral and octahedral; locally stalactitic; *color*: brass yellow on freshly broken surfaces.

quartz *abundance*: rare; *occurrence*: mudstone replacment; *crystallography*: microprismatic crystals; *color*: dark gray.

sphalerite *abundance*: uncommon; *occurrence*: cavity and fracture linings, replacement of limestone in contact with mudstone fill, growth in mudstone; *crystallography*: malformed tetrahedral crystals that are variably etched, sparry in fracture linings; *color*: dark brown (fig. 3.24).

Notes: In terms of the quality of mineral specimens it has produced, the Linwood Mine is arguably the best mineral-collecting locality in Iowa. Many museum-quality specimens of calcite, barite, and marcasite have been taken from the mine; many of these reside in the Anderson Museum at Cornell College and in the author's private collection. Unfortunately, mineral collecting in the mine is expressly prohibited because of underground mining hazards and resultant liability. Miners are permitted to extract minerals and sell them to collectors, but because miners lack sufficient time and adequate tools for proper collecting, mineral specimens obtained in this way are often damaged.

References: Garvin and Crawford 1992; Garvin 1995, unpublished data.

Mineral Creek Mines

Location: Allamakee County, about 16 kilometers north of Waukon; Sec. 13, T99N, R6W (fig. 3.13).

Host Rock Stratigraphy: Ordovician; Hager City Member, Oneota Formation.

Modes of Mineral Occurrence: dissolution-enlarged bedding plane and transverse fractures in carbonate rock; collapse breccia cements.

Minerals

anglesite *abundance*: uncommon; *occurrence*: coatings on galena; *crystallography*: microcrystalline; *color*: light to medium gray.

calcite *abundance*: abundant; *occurrence*: vug linings and sparry fracture fillings; *crystallography*: acute scalenohedra up to several centimeters in length that are deeply etched; *color*: white to light amber, frequently stained by limonite; *UV fluorescence*: nonfluorescent.

cerussite *abundance*: uncommon; *occurrence*: coatings on galena; *crystallography*: microcrystalline; *color*: light to medium gray, difficult to distinguish from anglesite in hand specimens.

dolomite *abundance*: abundant; *occurrence*: linings of small vugs and replacements of host carbonate rock; *crystallography*: generally microcrystalline; *color*: gray-brown but variably stained by limonite.

galena *abundance*: uncommon; *occurrence*: fracture linings and breccia cements; *crystallography*: single octahedral and cubo-crystals and subhedral clusters, individual crystals up to 1 centimeter across, rimmed by the oxidation products cerussite and anglesite; *color*: bright metallic gray on freshly broken surface; dull, dark gray on weathered surface (fig. 3.25).

goethite *abundance*: common; *occurrence*: vug and fracture linings; *crystallography*: microcrystalline masses, locally pseudomorphic after bladed marcasite and octahedral and radial colloform pyrite (the latter, the sooty pyrite of the formerly commercial zinc-lead district centered in southwestern Wisconsin); *color*: rust brown (fig. 3.25).

3.25. Distorted galena octahedra (light gray) on goethite pseudomorphic after marcasite. Mineral Creek Mines, Allamakee County. Specimen is 9 centimeters across. *Anderson Museum, Cornell College.*

hemimorphite *abundance*: common; *occurrence*: vug and fracture linings; *crystallography*: microcrystalline masses that replaced sphalerite; *color*: white to tan but variably stained by limonite, difficult to distinguish from smithsonite.

quartz *abundance*: abundant; *occurrence*: vug linings in and replacements of host carbonate rock; *crystallography*: generally microcrystalline chert and chalcedony, locally drusy; *color*: colorless to white but variably stained by limonite.

smithsonite *abundance*: common; *occurrence*: vug and fracture linings; *crystallography*: microcrystalline masses that replaced sphalerite; *color*: white to tan but variably stained by limonite.

sphalerite *abundance*: rare; *occurrence*: original sphalerite remains in a very few isolated locations in the mines; *crystallography*: sparry, banded, and interlayered with smithsonite and hemimorphite; *color*: light brown.

Notes: The Mineral Creek Mines were operated for a brief period during the 1850s. Two of the mines are still accessible. The original openings were small and have been made even smaller by flood-deposited silt. As a result, access to some areas of the mines requires squeezing through tight spaces. All minerals have experienced extensive weathering, and good-quality crystals are not likely to be found.

References: Calvin 1895; Leonard 1896; Heyl et al. 1959; Ludvigson 1976; Garvin 1982; Garvin et al. 1987.

Moscow Quarry

Location: Muscatine County, west of Moscow, on west side of Cedar River; Sec. 8, T78N, R2W (fig. 3.4).

Host Rock Stratigraphy: Middle Devonian; floor—Coggon Member, Otis Formation; rim—Solon Member, Little Cedar Formation; paleokarst fillings of Pennsylvanian age.

Modes of Mineral Occurrence: limestone—linings and fillings of dissolution cavities; fracture fillings; breccia cements; disseminations; paleokarst—disseminations in organic mudstone.

Minerals

calcite *abundance*: abundant; *occurrence*: cavity linings and fracture fillings; *crystallography*: several generations—early acute scalenohedral druse, followed by an obtuse rhombohedral generation, sometimes with brown coatings, followed by acute scalenohedron, frequently containing earlier rhombohedral phantoms that are highlighted by dustings of iron sulfide crystals; late crystals may reach 10 centimeters in length; rare rhombohedral twins; *color*: colorless, locally gray where containing abundant iron sulfide inclusions, root-beer brown; *UV fluorescence*: early spar—yellowish white; late scalenohedral—nonfluorescent (figs. 3.26, 3.27, pl. 11).

dolomite *abundance*: common; *occurrence*: scattered crystals in very vuggy host rock; *crystallography*: crystals are simple rhombohedra 3 millimeters or less across; *color*: white to cream-colored, locally coated with reddish brown iron oxide.

marcasite *abundance*: common; *occurrence*: chiefly in cavities; *crystallography*: as scattered blades on and within calcite, blades commonly show complex twinning and are locally in radiating groups, crystals are less than 1 centimeter in length; *color*: greenish gray on crystal surfaces, brass color on freshly broken surfaces.

pyrite *abundance*: common; *occurrence*: in cavities, encrustations on host rock, disseminations in paleokarst-filling mudstone and limestone; *crystallography*: scattered micro-octahedra and cubo-octahedra on and within

3.26. Calcite crystals with marcasite inclusions. Moscow Quarry, Muscatine County. Specimen is 6.5 centimeters across. *Anderson Museum, Cornell College.*

calcite, colloform encrustations; *color*: metallic brass yellow on freshly broken surfaces (fig. 3.27).

quartz *abundance*: uncommon; *crystallography*: ellipsoidal masses in excess of 10 centimeters in length, commonly cellular structure; *color*: colorless, very similar to quartz from Conklin Quarry.

sphalerite *abundance*: common; *occurrence*: as fracture fillings and cavity linings; *crystallography*: sparry masses and malformed tetrahedral crystals that are commonly etched; *color*: brown to red-brown, lustrous on freshly broken surfaces.

Notes: Calcite of excellent quality has been collected from the Moscow Quarry. The calcite is particularly attractive when exhibiting phantoms, which are highlighted by microscopic iron sulfide inclusions, and when containing perched radiating marcasite clusters. The best specimens are found in vugs in tough limestone and are difficult to remove without proper tools and much patience.

———

References: Horick 1974; Garvin, unpublished data.

Ollie Quarry

Location: Keokuk County, about 3.2 kilometers west of Ollie; Secs. 20 and 21, T74N, R11W (fig. 3.4).

Host Rock Stratigraphy: Mississippian; floor—Prospect Hill Formation; rim—Keokuk Formation.

Modes of Mineral Occurrence: cavity linings and fillings in limestone, commonly near areas of chert replacement; cavities generally less than 20 centimeters in length; replacement of limestone.

Minerals

barite *abundance*: uncommon; *occurrence*: late deposits on calcite and quartz; *crystallography*: small rosettes, bladed crystals up to 1 centimeter long; *color*: colorless, tan; *UV fluorescence*: pale creamy white.

calcite *abundance*: abundant; *occurrence*: cavity linings and fillings; *crystallography*: spar, modified acute scalenohedral crystals up to 4 centimeters long; *color*: colorless to white, locally may appear dark colored due to presence of sulfide inclusions; *UV fluorescence*: early spar—lemon yellow; late scalenohedral—pale pink.

fluorite *abundance*: uncommon; *occurrence*: cavity linings, with calcite; *crystallography*: cubic, crystals up to 5 centimeters across; *color*: translucent brown; *UV fluorescence*: bluish white.

marcasite *abundance*: uncommon; *occurrence*: scattered on quartz; *crystallography*: microbladed; *color*: dark greenish black on crystal surfaces.

3.27. Calcite crystals (white) on colloform pyrite (gray). Moscow Quarry, Muscatine County. Specimen is 11 centimeters across. *Anderson Museum, Cornell College.*

3.28. Colloform quartz. Ollie Quarry, Keokuk County. Specimen is 8 centimeters across. *Anderson Museum, Cornell College.*

millerite *abundance*: uncommon; *occurrence*: cavity fillings in chert and drusy quartz, inclusions in calcite, most common in upper stratigraphic units; *crystallography*: acicular, commonly in tufted masses and brushes; *color*: bronze yellow.

pyrite *abundance*: common; *occurrence*: nodular replacement in limestone, microcrystals scattered on drusy quartz and barite; *crystallography*: micro-crystalline, cubo-octahedral; *color*: reddish bronze on crystal surfaces.

quartz *abundance*: abundant; *occurrence*: cavity linings; *crystallography*: fine druse, crystals up to 1 centimeter long; locally colloform, spheres 3 to 4 millimeters across; *color*: white to dark gray (fig. 3.28).

sphalerite *abundance*: uncommon; *occurrence*: late void fillings on calcite, replacements in limestone; *crystallography*: sparry; *color*: medium to dark brown.

Notes: The Ollie Quarry is probably best known for its millerite; recent activity at the quarry has produced little good material.

References: Horick 1974; Garvin, unpublished data.

Peske Quarry

Location: Black Hawk County, on south side of U.S. Highway 20, near Raymond; Sec. 2, T88N, R12W (fig. 3.7).

Host Rock Stratigraphy: floor—Middle Devonian; Rapid Member—Little Cedar Formation; rim—Coralville Formation.

Modes of Mineral Occurrence: vugs, fracture fillings.

Minerals

barite *abundance*: uncommon; *occurrence*: perched on calcite as part of vug lining; *crystallography*: blades in radiating clusters, blade length up to 3 centimeters; *color*: nearly opaque white to light gray; *UV fluorescence*: dull creamy white to nonfluorescent (pl. 12).

calcite *abundance*: abundant; *occurrence*: vug linings, minor fracture fillings; *crystallography*: several generations of calcite are recognized; early rhombohedra, late acute scalenohedra; maximum size up to 3 centimeters long; *color*: early—white with light to dark brown coating, brown calcite locally exhibits purple iridescence; late—colorless to white; *UV fluorescence*: early—intense creamy white; late—dull pink to nonfluorescent (fig. 3.29).

fluorite *abundance*: uncommon; *occurrence*: vug linings; *crystallography*: simple cubes up to 2 centimeters across; *color*: opaque tan, pale yellow, root-beer brown; *UV fluorescence*: creamy white.

marcasite *abundance*: uncommon; *occurrence*: perched on and included in calcite; *crystallography*: small blades; *color*: pale brass yellow on freshly broken surfaces; gray green on oxidized surface.

3.29. Pseudocubic calcite crystals in vugs. Peske Quarry, Black Hawk County. Specimen is 13 centimeters across. *Anderson Museum, Cornell College.*

pyrite *abundance*: common; *occurrence*: single crystals and crusts on calcite, inclusions in calcite, minor fracture fillings; *crystallography*: the dominant crystal form is octahedron, maximum size 0.5 centimeter; *color*: brass yellow on freshly broken surfaces, bronze colored on oxidized surfaces.

sphalerite *abundance*: rare; *occurrence*: vug lining; *crystallography*: etched anhedral grains; *color*: dark brown.

Notes: The Peske Quarry is probably best known for its brown calcite and octahedral pyrite. Some of the brown rhombohedral (pseudocubic) calcite may be mistaken for fluorite. Effervescence in cold dilute hydrochloric acid identifies calcite. Barite clusters are excellent, but they are rare and are seldom found undamaged because of blasting. The host rock is unusually hard and relatively fracture free; hence, it is often difficult to remove the vug-containing minerals.

———

Reference: Garvin, unpublished data.

Pint's Quarry

Location: Black Hawk County, east of Waterloo at Raymond; Sec. 36, T89N, R12W (fig. 3.7).

Host Rock Stratigraphy: floor—Middle Devonian; Rapid Member—Little Cedar Formation; rim—Coralville Formation.

Modes of Mineral Occurrence: vug linings; in the Rapid Member vugs were formed through complete or partial replacement of colonial corals.

Minerals

barite *abundance*: common; *crystallography*: tabular to bladed crystals that commonly occur in radiating clusters and in subparallel stacks, individual crystals up to 8 centimeters across, rare microprismatic crystals perched on calcite; *color*: colorless, white, frequently stained by limonite; *UV fluorescence*: nonfluorescent (fig. 3.30).

calcite *abundance*: abundant; *crystallography*: early acute scalenohedral crystals up to 10 centimeters in length, rare acute rhombohedra up to 3 centimeters across, late calcite is sparry and locally rhombohedral; *color*: colorless, white, pale yellow, rarely pink, locally brown organic coatings; *UV fluorescence*: early—creamy white; late scalenohedral—dull to bright pink (pl. 13).

3.30. "Stack" of bladed barite crystals. Pint's Quarry, Black Hawk County. Specimen is 9 centimeters across. *Anderson Museum, Cornell College.*

fluorite *abundance*: common; *crystallography*: cubic, crystals range in size from a few millimeters to 3 centimeters; *color*: very pale yellow to dark brown, rarely purple; *UV fluorescence*: all varieties (including purple) intense whitish yellow to creamy white (pl. 14).

galena *abundance*: rare; *crystallography*: sparry; *color*: bright metallic gray.

gypsum *abundance*: rare; *occurrence*: intergrown with calcite; *crystallography*: microcrystalline; *color*: light gray.

marcasite *abundance*: common; *crystallography*: blades and prisms up to 3 centimeters in length, tabular clusters, complex twins and intergrowths, locally colloform masses; *color*: pale brass yellow on freshly broken surfaces, gray green to brown on oxidized surfaces; more abundant in Coralville Formation than in Rapid Member host rocks.

pyrite *abundance*: abundant; *crystallography*: single crystals and crystal clusters, commonly scattered on and as inclusions in calcite; octahedral and cubo-octahedral forms most common; individual crystal size up to 3 centimeters across; *color*: due to coatings by iridescent oxidation films, colors range throughout the spectrum, with red, blue, and green the most common; the thickness of the film determines the color (fig. 3.31).

3.31. Pyrite clusters on dolostone. Pint's Quarry, Black Hawk County. Specimen is 9.5 centimeters across. *Anderson Museum, Cornell College.*

quartz *abundance*: abundant; *occurrence*: pseudomorphic replacement of colonial coral, microcrystalline to prismatic, prismatic crystals up to 1 centimeter in length; *color*: colorless to white.

sphalerite *abundance*: uncommon; *crystallography*: small (< 1 centimeter), poorly formed tetrahedra; *color*: dark brown.

Notes: Except for the geode localities, Pint's Quarry boasts the greatest variety of minerals anywhere in Iowa. Nine different minerals have been reported from vug linings. Fine specimens of fluorite, calcite, marcasite, and pyrite have been collected. Pyrite is particularly attractive because of the iridescent coloration. Regrettably, the quarry has been closed for several years, and the best areas for collecting (in the lower part of the quarry) are flooded.

References: Menzel and Pratt 1968; Lin 1978; Anderson and Garvin 1984; Anderson and Stinchfield 1989; Garvin, unpublished data.

Postville Quarry

Location: southern Allamakee County, on State Highway 51, about 3.2 kilometers north of Postville; Sec. 16, T96N, R6W (fig. 3.13).

Host Rock Stratigraphy: Ordovician; floor—Wise Lake Formation; rim—Elgin Member, Maquoketa Formation; most mineralization occurs at the contact between the Wise Lake Formation and the Elgin Member; also pyrite in shales and in phosphatic material at the base of the Elgin Member.

Modes of Mineral Occurrence: small ellipsoidal vugs; replacements of phosphatic shale; nodules and disseminations in shale; replacement of limestone.

Minerals

barite *abundance*: uncommon; *occurrence*: on calcite and fluorite; *crystallography*: small (< 2 millimeters) prismatic crystals; *color*: colorless; *UV fluorescence*: nonfluorescent.

bitumen *abundance*: rare; *occurrence*: tiny black spheres of what appears to be bitumen occur on surfaces of calcite and fluorite.

calcite *abundance*: abundant; *occurrence*: vug linings; *crystallography*: at least two generations identified—early acute scalenohedra, late obtuse rhombohedra; *color*: colorless, white, some late calcite is pale yellow; *UV fluorescence*: early calcite—creamy white; later varieties—nonfluorescent (fig. 3.32).

fluorite *abundance*: common; *occurrence*: vug linings; *crystallography*: simple cubes, minor modification by trapezohedron, maximum size 0.5 centimeter; *color*: yellow, colorless, reddish to bluish purple; pronounced zon-

3.32. Calcite crystals in vug. Postville Quarry, Allamakee County. Vug is 7 centimeters across. *Anderson Museum, Cornell College.*

ing, with yellow cores and purple rims the most common; *UV fluorescence*: yellow varieties—dull yellow to bright yellow-white; purple varieties—nonfluorescent.

gypsum *abundance*: uncommon; *occurrence*: fills empty space in cavities; *crystallography*: bladed to sparry, generally strongly etched; *color*: colorless to white, locally pale blue.

marcasite *abundance*: rare; *occurrence*: vug linings, locally projecting from pyrite cubes; *crystallography*: microscopic blades, less than 1 millimeter in length, extremely fragile; *color*: dark gray-green.

pyrite *abundance*: common; *occurrence*: perched on and included in calcite, single crystals and nodular clusters in shale, replacement of fossils in phosphatic layer; *crystallography*: cubes and cubo-octahedra; maximum size a few millimeters across in vugs, a centimeter or more across in shale; *color*: brass yellow on freshly broken surfaces, bronze colored on oxidized surfaces, iridescent coloration in phosphatic shales.

siderite *abundance*: rare; *occurrence*: scattered on calcite and fluorite; *crystallography*: simple rhombohedron; *color*: red brown, under high magnification appears as red patches in uncolored matrix.

sphalerite *abundance*: rare; *occurrence*: along walls of vugs, replacing host; *crystallography*: poorly formed tetrahedra and subhedral clusters; *color*: dark brown.

Notes: The fluorite cubes, though small, are attractive. Specimens are probably best for thumbnails or micromounts. The collecting area in the Postville Quarry is limited. This deposit is very similar to the Volga deposit.

———

References: Horick 1974; Garvin, unpublished data.

Robins Quarry

Location: Linn County, north of Cedar Rapids at Robins; Sec. 21, T84N, R7W (fig. 3.7).

Host Rock Stratigraphy: Middle Devonian; floor—Spring Grove Member, Pinicon Ridge Formation; rim—Solon Member, Little Cedar Formation; paleokarst fillings of undetermined Pennsylvanian age.

Modes of Mineral Occurrence: limestone—dissolution-enlarged fracture linings and fillings; minor vug linings; paleokarst (mudstones and sandstones)—nodules and disseminations in organic-rich mudstone; replacement of woody plant material.

3.33. Zoned calcite crystals. Robins Quarry, Linn County. Specimen is 13 centimeters across. *Anderson Museum, Cornell College.*

Minerals

calcite *abundance*: common; *occurrence*: fracture linings and fillings, a single large fissure fill about 2 meters thick, vug linings in a vuggy horizon of the Spring Grove Member; *crystallography*: fissure fill calcite—sparry, with individual cleavage rhombs up to 16 centimeters across; vug-lining calcite—zoned acute rhombohedra, with crystals up to 1 centimeter across; fracture-lining calcite—modified acute scalenohedral, with crystals up to 12 centimeters in length; *color*: fissure fill calcite—white with patches of tan; vug-lining calcite—colorless; fracture lining calcite—colorless to white, locally may appear dark gray due to inclusions of FeS_2; *UV fluorescence*: Spring Grove–hosted rhombohedral variety—whitish yellow; all other varieties dull or nonfluorescent (figs. 1.1, 3.33).

marcasite *abundance*: uncommon; *occurrence*: small replacement nodules in limestone, minor vug linings; *crystallography*: crystals are typical blades and wedges less than 0.5 centimeter in length; *color*: pale brass yellow on freshly broken surfaces, metallic greenish gray on oxidized surfaces.

pyrite *abundance*: common; *occurrence*: fracture fillings in limestone, nodules in organic-rich paleokarst-filling mudstone, minor replacement of carbonized wood, cave linings in contact with mudstone; *crystallography*: fracture-filling pyrite—microcrystalline; nodular pyrite—intergrown cubes up to 3 centimeters across; cave linings—microcrystalline, exhibiting a stalactitic appearance; *color*: brass yellow on freshly broken surfaces, rust colored on oxidized surfaces (fig. 3.34).

3.34. Pyrite cubes. Robins Quarry, Linn County. Specimen is 4 centimeters across. *Anderson Museum, Cornell College.*

quartz *abundance*: uncommon; *occurrence*: replacement of limestone; *crystallography*: ellipsoidal masses, locally exhibiting cellular structure; *color*: white to light gray; very similar to occurrence at Conklin Quarry.

sphalerite *abundance*: uncommon; *occurrence*: fracture fillings and replacements of limestone; *crystallography*: sparry; *color*: medium brown.

Notes: The Robins Quarry is better known for its fossils than for its minerals. Excellent brachiopods and silicified corals of several varieties can be found here. Large, single pyrite crystals and crystal clusters occasionally are found in paleokarst-filling mudstones. The large sparry calcite fissure is no longer exposed. Recent stripping of regolith along the north side of the quarry exposed a localized concentration of large, zoned calcite crystals, the first I have seen at Robins.

Reference: Garvin, unpublished data.

Sperry Gypsum Mine

Location: Des Moines County, 6.4 kilometers southeast of Mediapolis at Sperry; NW 1/4, Sec. 3, T71N, R3W (fig. 3.4).

Host Rock Stratigraphy: gypsum and anhydrite beds constitute part of the Middle Devonian Spring Grove Formation.

Mode of Geological Occurrence: beds and lenses that commonly exhibit laminations and are locally distorted due to stresses related to hydration of anhydrite; localized nodular masses; veins, vugs, and replacements containing secondary minerals.

Minerals

anhydrite *abundance*: abundant; *occurrence*: lenses, in part interlayered with dolomite; *crystallography*: fine-grained, massive; *color*: pale blue.

celestite *abundance*: uncommon; *occurrence*: veins and vug linings, replacements of anhydrite; *crystallography*: tabular to bladed crystals and crystalline masses, individual crystals several centimeters in length; *color*: colorless to pale blue (fig. 3.35).

dolomite *abundance*: abundant; *occurrence*: interlaminated with gypsum and anhydrite; *crystallography*: fine-grained massive, locally euhedral rhombs; *color*: brown, black.

gypsum *abundance*: abundant; *occurrence*: beds, lenses; secondary replacements of and veins cutting across anhydrite; vug linings; *crystallography*: fine-grained, massive; microblades and equant crystals; vug-lining crystals prismatic, up to several centimeters long; *color*: fine-grained—white to gray; vug-lining crystals—transparent to translucent colorless.

quartz *abundance*: common; *occurrence*: ovoid-shaped nodular masses at least 25 centimeters in diameter and 40 centimeters long, concentrated in

3.35. Bladed celestite crystals. Sperry Gypsum Mine, Des Moines County. Specimen is 6 centimeters across. *Anderson Museum, Cornell College.*

upper part of gypsum bed about 1 meter below roof of mine; *crystallography*: fine-grained, massive; *color*: white to gray.

Notes: The Sperry Mine is well known for its celestite and selenitic gypsum crystals, but crystallized areas are very localized. As with other underground mines in Iowa, mineral collecting is generally prohibited.

———

References: Dorheim et al. 1972; Sendlein 1973.

Volga Exposure

Location: Clayton County, on Volga River, just upstream from abandoned dam site at Volga; Secs. 3 and 10, T92N, R6W (fig. 3.13).

Host Rock Stratigraphy: Ordovician; Elgin Member, Maquoketa Shale (phosphatic zone) and uppermost Dubuque Member, Galena Formation.

Modes of Mineral Occurrence: linings of small (1 to 9 centimeters across) vugs; minor fillings of fractures connecting vugs.

Minerals

bitumen *abundance*: rare; *occurrence*: tiny lustrous black globules of what appears to be hydrocarbon material appear as specks on some crystal surfaces.

calcite *abundance*: abundant; *occurrence*: vug linings, fracture fillings; *crystallography*: stage I—sparry; stage II—scalenohedra, generally 5 to 15 millimeters long but as long as 8 centimeters; rare rhombohedral twins; stage III—prism terminated by rhombohedra, 1 to 2 millimeters long; *color*: stage I—colorless; stage II—milky white; stage III—colorless to milky white; *UV fluorescence*: stage I—whitish; stages II and III—nonfluorescent.

fluorite *abundance*: common; *occurrence*: vug linings, scattered crystals on stage I calcite; *crystallography*: simple cubes up to 5 millimeters across, granules; *color*: colorless to pale pink, yellow, purple, colorless from oldest to youngest, crystals are strongly zoned; *UV fluorescence*: yellow fluorite exhibits whitish fluorescence, other colors nonfluorescent.

marcasite *abundance*: uncommon; *occurrence*: does not occur in vugs but as disseminations and fossil replacements in upper phosphorite bed, about a half meter above the fluorite-bearing layer.

Plate 1. Calcite. Buffalo Quarry, Scott County. Six centimeters high. *Anderson Museum, Cornell College.*

Plate 2. Smoky quartz crystal and goethite needles on calcite. Marion County. Twelve centimeters across. *David Malm collection.*

Plate 3. Pyrite and marcasite crystals on calcite. Conklin Quarry, Johnson County. Twelve centimeters high. *Anderson Museum, Cornell College.*

Plate 4. Calcite basal twin on limestone. Conklin Quarry, Johnson County. Seven centimeters across. *Anderson Museum, Cornell College.*

Plate 5. Quartz on agate. Keswick Quarry, Keokuk County. Twenty centimeters across. *David Malm collection.*

Plate 6. Plumose barite. Linwood Mine, Scott County. Six centimeters across. *Anderson Museum, Cornell College.*

Plate 7. Iridescent marcasite. Linwood Mine, Scott County. Nine centimeters across. *Anderson Museum, Cornell College.*

Plate 8. Calcite and marcasite boxwork. Linwood Mine, Scott County. Eight centimeters across. *Anderson Museum, Cornell College.*

Plate 9. Barite. Linwood Mine, Scott County. Nine centimeters across. *Anderson Museum, Cornell College.*

Plate 10. Calcite with marcasite inclusions. Linwood Mine, Scott County. Nine centimeters high. *Anderson Museum, Cornell College.*

Plate 11. Calcite with marcasite inclusions. Moscow Quarry, Muscatine County. Five centimeters across. *Anderson Museum, Cornell College.*

Plate 12. Barite rosette. Peske Quarry, Black Hawk County. Rosette is five centimeters across. *Anderson Museum, Cornell College.*

Plate 13. Calcite. Pint's Quarry, Black Hawk County. Seven centimeters high. *Anderson Museum, Cornell College.*

Plate 14. Fluorite on calcite. Pint's Quarry, Black Hawk County. Ten centimeters across. *David Malm collection.*

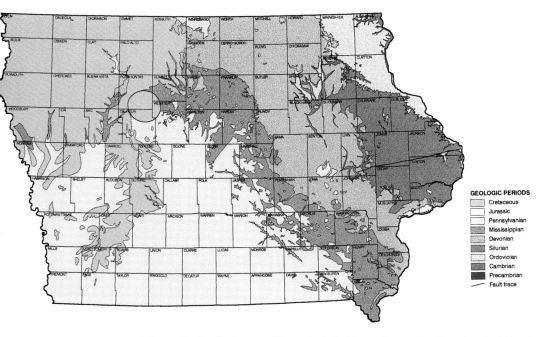

Plate 15. Geologic map of Iowa. *Geological Survey Bureau, Iowa Department of Natural Resources.*

Plate 16. The George Washington bottle, a sand painting by Andrew Clemens. The bottle is 11.5 centimeters in diameter and 30.5 centimeters high. *State Historical Society of Iowa, Iowa City.*

pyrite *abundance*: uncommon; *occurrence*: crusts on stage I calcite, inclusions in later calcite; *crystallography*: cubes, pyritohedra, octahedra, and combinations; crystals are millimeter-sized; some crystals grown together in beadlike rods a few millimeters long; *color*: brass yellow but tarnished due to oxidation.

sphalerite *abundance*: uncommon; *occurrence*: growing on cavity walls, intergrown with stage I calcite; *crystallography*: crystal forms not readily apparent, rare spinel twins, crystals up to 5 millimeters across; *color*: very dark brown, thin fragments smoky yellow.

Notes: The mineral deposit at the Volga Exposure bears a striking similarity to the deposit at the Postville Quarry, except that it apparently lacks barite, gypsum, and siderite. This exposure is normally submerged by the Volga River; thus, the best opportunities for collecting occur when water levels are very low. Because of the small size of crystals and vugs, the exposure is probably best for micromounts or thumbnail specimens. The similarity between the Volga and Postville occurrences suggests that, wherever phosphatized basal Maquoketa Shale is exposed in northeast Iowa, mineralization might be present.

Reference: Brown 1967.

Waterloo South Mine

Location: Black Hawk County, about 16 kilometers south of Waterloo; Sec. 18, T87N, R12W (fig. 3.7).

Host Rock Stratigraphy: floor—Middle Devonian; Rapid Member—Little Cedar Formation, rim—Coralville Formation.

Modes of Mineral Occurrence: vug linings, breccia cements, fracture linings; the mine contains prominent dissolution-enlarged fissures, which have subsequently filled with mud; the fissures are generally barren of mineralization.

Minerals

barite *abundance*: uncommon; *occurrence*: vug linings and breccia cements; *crystallography*: blades, commonly in radiating clusters; *color*: nearly opaque white; *UV fluorescence*: nonfluorescent.

calcite *abundance*: abundant; *occurrence*: vugs and bedding plane and transverse fractures; *crystallography*: acute scalenohedron and prism, modified

3.36. Fluorite crystals on calcite. Waterloo South Mine, Black Hawk County. Specimen is 9 centimeters across. *Paul Garvin collection.*

by rhombohedron; some late calcite is sparry; crystals up to several centimeters in length; *color*: early—brownish yellow; late—colorless to white; *UV fluorescence*: early—creamy white; late—dull pink to nonfluorescent (fig. 3.36).

chalcopyrite *abundance*: rare; *occurrence*: perched on calcite; *crystallography*: microscopic pseudotetrahedral crystals; *color*: metallic yellow on freshly broken surfaces; reddish yellow on oxidized surface.

fluorite *abundance*: common; *occurrence*: vug linings, on calcite; *crystallography*: simple cubes, locally with curved faces; crystals may reach 2 centimeters across; *color*: pale yellow to root-beer brown, locally nearly opaque tan; *UV fluorescence*: opaque variety—intense yellowish white; root-beer variety—yellowish gray (fig. 3.35).

marcasite *abundance*: common; *occurrence*: commonly perched on pyrite crystals; *crystallography*: thin blades and capillary crystals; *color*: pale brass yellow on freshly broken surfaces, greenish gray on oxidized surfaces.

pyrite *abundance*: common; *occurrence*: crusts on fracture walls, breccia cement, inclusions in calcite; *crystallography*: combinations of octahedron, cube, and pyritohedron; maximum size is a few millimeters across; *color*: brass yellow on freshly broken surfaces but may exhibit other colors due to iridescent oxidation films.

quartz *abundance*: rare; *occurrence*: replacement of tabulate coral; *crystallography*: fine druse, crystals doubly terminated; *color*: colorless.

sphalerite *abundance*: rare; *occurrence*: vug lining; *crystallography*: etched, malformed tetrahedra; *color*: dark brown.

Notes: The mineralogy at the Waterloo South Mine is quite similar to that at Pint's Quarry, except that there is less iron sulfide and barite at Waterloo South. The best specimens, including all the fluorite, have come from the mine, which is now closed and flooded. An adjacent quarry, still in operation, exposes rocks of the Coralville Formation, which are not nearly as well mineralized.

References: Anderson and Garvin 1984; Garvin, unpublished data.

Miscellaneous Occurrences

The mineral occurrences chosen for inclusion in this book are those that are well known to mineral scientists and collectors. Because of their chemical and physical nature, carbonate rocks are favorable sites for the formation of epigenetic calcite and sulfide minerals, especially where the rocks are in close proximity to mudstones or shales that are rich in organic matter. Thus, close inspection of the rocks exposed at any limestone or dolostone quarry, especially in eastern Iowa, may reveal the presence of these minerals. Minor occurrences reported in the literature but not described in this book include the Fairbank, Ferguson, and Twin Springs quarries; the G. Huber Farm (Spry and Kutz 1988); and the Anamosa, Dyersville, Lytle Creek, Mineral Creek, and other (unnamed) ore prospects in northeast and east-central Iowa (Heyl et al. 1959). Galpin (1922) reported finding geodes at Rockford (Floyd County); Udden (1898) lists several minerals in an Iowa Geological Survey report on Muscatine County; and Kuntze (1899) claims to have identified the mineral quenstedite near Montpelier in Muscatine County. Horick (1974) lists additional minor occurrences of specific minerals.

4. Iowa's Mineral Industries

Iowa has produced a variety of mineral commodities and has been a major producer of some. For some minerals, the periods of production were relatively short, and the mines have long since been abandoned; for others, production continues to the present. Iowa's mineral industries make a significant contribution to the state's economy (Anderson 1979). This chapter will summarize the history and current status of the following industries: clay, coal, gold, gypsum, iron, lead and zinc, limestone, peat, petroleum and natural gas, sand and gravel, and silica. Some may question the inclusion of petroleum, natural gas, and peat because, in a technical sense, they are not minerals. Neither are rock gypsum, limestone, clay, or sand and gravel. However, as discussed in chapter 1, for industrial and economic purposes, they are considered mineral resources, and it is entirely appropriate to include them in a book on Iowa minerals.

Iowa has enjoyed a long history of rock and mineral production and use. For millennia prior to Euroamerican settlement of Iowa, Native Americans used rocks and minerals for implements, weaponry, and adornment. They also used rocks and minerals in ceremonies. Iowa chert was widely used in the production of chipped-stone tools. Chert-bearing Paleozoic carbonate rocks are widespread in all but the south-central and northwest areas of the state. Color patterns, texture, and fossils present in chert artifacts are used to identify their sources (Morrow 1994). Medium- and low-quality cherts were used mainly near their sources. High-quality material, like Burlington Chert, was used over a much wider area. Sioux Quartzite, from the northwest corner of the state, was used chiefly as grinding implements. Clay was extracted from alluvial and terrace deposits for pottery making. The sources of the clay used in the manufacture of pottery at specific archaeological sites have not been determined. Earthy hematite and limonite (ochers) were ground and used as pigments. Basalt, granite, quartzite, and other durable igneous and metamorphic rocks in glacial till were shaped for use as grinding and pounding implements. Gypsum may

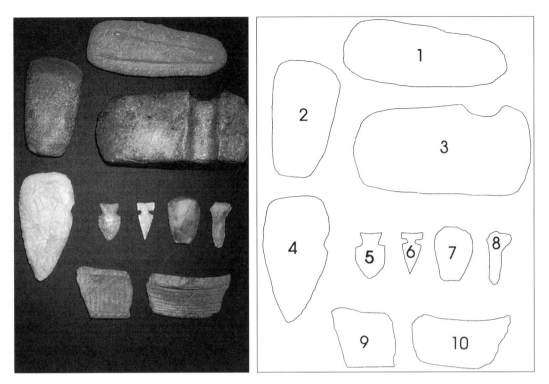

4.1. Native American artifacts made of materials found in Iowa. (1) abrader (sandstone); (2) celt (basalt); (3) three-quarter groove axe (basalt); (4) blade; (5) and (6) projectile points; (7) scraper (chert); (8) drill (chert); (9) and (10) pottery fragments (clay). Axe is 13 centimeters long. *Artifacts provided by the Office of the State Archaeologist, University of Iowa.*

have been used for medicinal purposes. Galena was used for beads and pigment (William Green, personal communication). The dates of first use of lead and coal are not known, but evidence suggests that they might go back at least 800 years (Grant 1979; Bettis and Green, 1993). There is some evidence that lignite coal from the Big Sioux area near Plymouth was used by the Ioways for fuel (Keyes 1912a) (fig. 4.1).

Clay

The use of clay in Iowa can be traced back to early Euroamerican settlement. Hall and Whitney (1858), in their survey of Iowa, reported clay suitable for brickmaking in "almost every district of the State." They also reported the existence of "fireclay" underlying coal beds in the coal measures of central and southern Iowa and cited its use in the manufacture of pottery. Charles White (1870) recognized the existence of surface ("drift") clays and bedrock shales, and he believed that Iowa drift clays were ideal for the manufacture of brick because they contained, in addition to clay, sand (necessary for strength and uniform shrinkage during firing) and

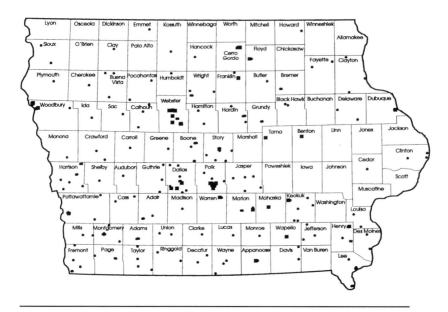

4.2. Producers of clay products in Iowa in 1902 (circles) and 1940 (squares).
Data from Beyer and Williams 1904; Gwynne 1943.

lime (important as a flux to insure complete fusibility). The best pottery clays, in his opinion, were to be found in the underclays of Iowa's coal measures. At the time of his report in 1870, early potteries in Iowa included those at Eldora (Hardin County), Fairport (Muscatine County), Des Moines (Polk County), Vernon (Van Buren County), Danville (Des Moines County), and Boonsboro (Boone County). He did not consider coal underclays suitable for firebrick because they contained too much lime and other fluxes, which lower the temperature of fusibility (White 1870).

By the early 1900s the clay industry in Iowa had expanded greatly. At least 380 producers of clay products operated in eighty-nine Iowa counties (McKay 1992) (fig. 4.2). Early uses included drainage tile, brick (common, facing, and paving), sewer pipe, pottery, and stoneware, and burnt clay was used for railroad ballast. Surface clays were obtained from alluvial, glacial, and loess sources and their weathering residua. Bedrock clays came primarily from the Ordovician Maquoketa Formation (northeast and east-central Iowa), the Devonian Juniper Hill Member of the Lime Creek Formation (northern Iowa), the rocks of the Pennsylvanian Des Moines Series (southern and central Iowa), and Cretaceous rocks (northwest Iowa).

Extraction

The method of clay extraction depends on the physical character of the clay (e.g., degree of induration), the depth of overburden, and the stratigraphic relationship of clay to adjacent rock. Surface clays at first were removed with pick and shovel and later with plow and scraper. Well-indurated shales were quarried with steam shovels. Where necessary, blasting was done. Where the depth of overburden was too great for surface removal, room and pillar and longwall underground mining methods were employed. Mining was used only in those cases where the clay was highly desirable and therefore brought a suitably high price (Beyer and Williams 1904).

Hauling

Surface clays were hauled to the processing plant by wheelbarrow, provided the distance was short; otherwise, a horse-drawn two-wheeled cart was most commonly used. In some cases soft surface clays were removed and hauled by means of wheeled scrapers. Where the demand for clay was great enough, rail spurs were brought to the pit, and the clay was hauled by train to the processing plant (Beyer and Williams 1904).

Preparation of Clay Products

Basically, preparation of the raw material involved crushing and grinding (if the material was well indurated) and mixing (required because the raw material was seldom uniform in its consistency). Crushing could be wet or dry. The product was generally screened to assure uniform size (Beyer and Williams 1904).

In the manufacture of bricks, plastic wet clay was first pressed into molds (e.g., brick molds) by hand or machine. By hand, an experienced brick molder could turn out about 1,600 bricks in six hours. In contrast, a "stiff mud" machine could produce 8,000 to 10,000 bricks per hour. Next, molded clay forms were dried in the open air or on racks in heated dryers. Frequently, dried forms were re-pressed to improve form, to increase density and strength, and to imprint designs. Re-pressing was commonly used in the production of paving brick. The "green" clay forms were then fired (burnt) in wood- or coal-fired kilns. Firing drove off any remaining water and produced structural/chemical changes that gave the finished product strength and durability (Beyer and Williams 1904) (fig. 4.3). More recently, oil and natural gas have been used as heat sources.

Production

After 1900 the number of clay producers began to decline, but for a time the tonnage and value of the clay continued to increase. Larger producers and the increased value of clay products were the reason. In 1920 Iowa led the nation in the production of drain tile (McKay 1992). By 1940 the number of producers had shrunk, and their products were exclusively brick, drain tile, building tile, and sewer pipe (fig. 4.2). The decline in the industry reflects the change from using relatively inexpensive-to-mine surface clays to more expensive shales (Gwynne 1943). It also results from

4.3. Early production of clay products. Iowa Pipe and Tile Works, Des Moines, ca. 1896. *Calvin Photographic Collection, Department of Geology, University of Iowa.*

the conversion from paving brick to other road-surfacing materials. With the closing of the long-standing Rockford Brick and Tile plant in 1992, only three clay producers remain in Iowa. These firms are in Dallas and Woodbury counties, and they produce only facing brick (McKay 1992) (fig. 4.4).

Early efforts were made to identify and develop Iowa sources of refractory clay. Although underclay, which is frequently used as a source, is abundant beneath the coals of Iowa, the composition of the underclay is such that its fusibility is too low to meet heat-tolerance requirements (Galpin 1925).

Coal

The coal mining industry in Iowa enjoyed a long and productive history, exceeded in length and value of product only by the limestone industry. During its zenith it was second only to agriculture. Coal has been mined from thirty-four central and southern Iowa counties (Keyes 1894a; Hinds 1909; Howes 1992). The first report of coal in Iowa was in 1817, when early explorers observed many fragments of coal on sandbars in the

Des Moines River (Van der Zee 1915). During an 1836 investigation of the lands acquired in the Black Hawk Purchase, Albert Lea reported coal along the banks of the Des Moines River near its confluence with the Raccoon River (Schwieder 1983).

The first reported use of coal in Iowa was the fueling of a blacksmith's forge at Fort Des Moines in 1840. In that same year Samuel Knight opened a coal mine at the village of Farmington in Van Buren County (Schwieder 1983). Iowa coal was used primarily for residential heating and in forges and potteries. The development of steamboat traffic on the Des Moines River in the 1840s and 1850s increased demand for Iowa coal. But demand was seasonal, and the mines generally closed during the summer months—until the arrival of the railroads in Iowa. By 1860 Iowa claimed

Production of Clay and Clay Products in Iowa

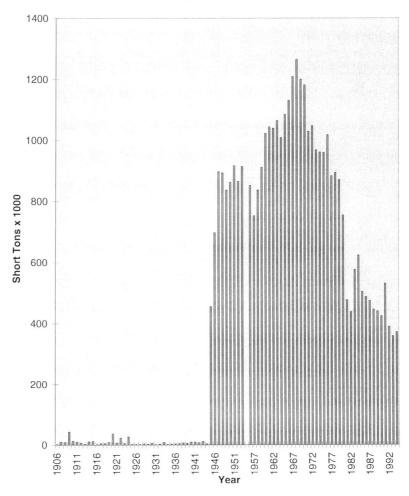

4.4. Production of clay products in Iowa. *Data from the Geological Survey Bureau, Iowa Department of Natural Resources and the United States Bureau of Mines.*

800 kilometers of track. Demand for coal increased dramatically, and mines began to operate on a year-round basis. By 1876 the railroads had reached Council Bluffs, and Iowa was the leading coal-producing state west of the Mississippi River and fifth in the nation (Howes 1992).

Mining Methods

The earliest coal workings were diggings from coal banks, which were outcrops along the Des Moines and other rivers. Coal was hauled by wheelbarrow and dumped into wagons. Thirty bushels was considered a load. As diggings extended into the banks, wooden timbers were placed to support roofs, and drift mining began. The discovery of coal veins that were too deep for drift mining gave rise to slope mining and shaft mining. In slope mining, inclined entries were made from the surface to the seam, at angles that permitted coal to be hauled by animal- or human-powered cars to the surface. For deeper seams, the slope distance from the surface to the seam was a kilometer or more. If seams were more than 30 meters deep, vertical shafts were driven. Coal was raised in buckets by men or by horse-powered gins and later by steam-powered hoists (fig. 4.5). In some counties mines were worked on more than one level; this was particularly true in Polk County, where levels were referred to as the first, second, or third vein. In the early 1870s at the Happy Hollow Mine in Appanoose County coal was first extracted from a drift, but about 70 meters back from

4.5. Pekay Coal Mine, Mahaska County. *Courtesy of the Calvin Photographic Collection, Department of Geology, University of Iowa.*

the entrance the seam began to pinch out. In the process of drilling a water well on the property, a lower seam was found. A pit was dug on the drift floor, and the same seam was encountered about 5 meters below the pit floor. The pit was enlarged, and a steep grade was constructed to connect the two levels. A mine locomotive was purchased, the first used in Iowa, for the purpose of hauling coal from the lower level. On its initial run down the incline the new engine malfunctioned and had to be hauled out by ropes. It spent the remainder of its days performing tasks on the surface (Lees 1909).

Underground coal was extracted by either the longwall or the room and pillar method. In longwall mining the coal seam was undercut at the base and a half meter or more of material was removed. Within twenty-four hours the weight of the coal normally caused the coal to break away from the ceiling into the cavity created by undercutting. Where necessary, coal was wedged down or blasted loose. The material from the undercut was piled from floor to ceiling behind the working face to provide ceiling support. Ceilings were also supported by cribs made of wooden timbers and packed with undercut material. In this way all the coal could be mined out. Longwall workings typically fanned out from the shaft (Schwieder and Kraemer 1973) (fig. 4.6).

In room and pillar mining two parallel tunnels were run from the shaft or slope bottom. At intervals of about 100 meters cross entries were driven at right angles to the tunnels. The rooms were opened both ways from the cross entries, which made them parallel to the main tunnels. A room was about 9 meters wide and when finished was about 40 meters long; each room was separated from the adjacent room by a wall of unmined coal about 3 meters thick (Schwieder and Kraemer 1973) (fig. 4.7). Pillars were often shaved or removed as the mine was being abandoned. By this method only about 50 percent of the coal could be mined, but mining was safer (Stolp and Deluca 1976).

Mining Life

The earliest coal mines in Iowa belonged to local landowners. With the establishment of railroads, railroad companies developed the practice of leasing coal lands and operating mines as subsidiaries. These mines became known as captive mines. One of the largest subsidiaries was the Consolidation Coal Company, owned by the Chicago and North Western Railroad. Its mines at Buxton became the largest captive mines in the state (Schwieder 1983).

As the number and size of mines in Iowa increased, the demand for miners also grew. Miners coming from the eastern United States brought town names, such as Pittsburgh and Cincinnati, to the coalfields. The mines at Buxton employed a large number of African Americans (Schwieder 1983; Howes 1992). Increasing demand for miners was met by foreign-born

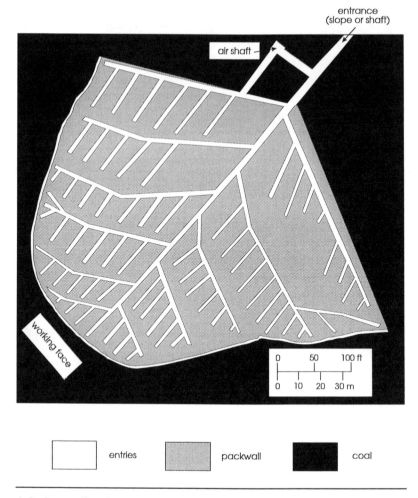

entrance
(slope or shaft)

air shaft

working face

0	50	100 ft
0	10	20	30 m

entries   packwall   coal

4.6. Longwall coal mine plan. *After Schwieder (1983).*

immigrants. Miners from England and Wales were considered by many to be the best because they were seasoned with experience from English coal mines. Coal miners also came from Scotland, Sweden, Italy, and Croatia and other countries in eastern Europe. After 1900 Italians dominated the foreign-born miner population (Schwieder 1982).

Mining companies employing large numbers of workers frequently built company towns. Typically, the town consisted of straight rows of houses on both sides of a main street (fig. 4.8). The central building was the company store, where miners and their families could obtain groceries and other items. The town of Angus in Boone County, with a population of 3,500, became the largest mining company town in the state (Schwieder 1983). Because the average life of an Iowa coal mine was no more than ten years, company towns generally were poorly maintained. Miners were often paid in scrip, which could be used only at the company store. When noncompany stores appeared, miners were urged to trade only at the company store or risk being fired. Miners often bought on credit, and some "owed their souls to the company store." Most company towns have disappeared, though Beacon in Mahaska County and Hiteman in Monroe County still remain as small communities (Howes 1992).

The coal miner in Iowa led a hard life. Coal seams averaged a meter thick, and the floor would be lowered to allow mules and pit cars access to the workings. Generally this meant that miners could not stand up (fig. 4.9). In Scott County, where the seams were even thinner, tunnels were too low for animals, thus miners had to crouch in order to push the pit cars to the hoist (Lees 1909). In southern Iowa, where seam thickness averaged less than a meter, miners had to dig out coal while lying on their

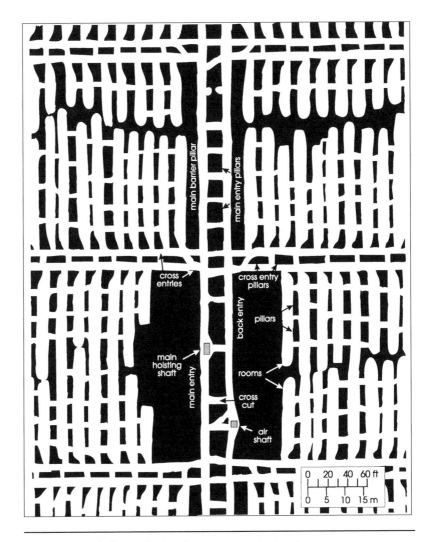

4.7. Room and pillar coal mine plan. *After Schwieder (1983).*

4.8. Coal company town, 1900–1910. *State Historical Society of Iowa, Iowa City.*

sides. In addition to enduring cramped working conditions, miners had to pay for their tools and transportation. A company town that sprang up near the mine shaft became progressively separated from the mine as new shafts were built to follow the seam. As a result, miners who could not afford transportation walked as much as 16 kilometers to work—summer and winter. Miners were paid not by the hour but by the weight of coal hauled from the mine. Time spent shoring up timbers, lowering (brushing) the floor, or undercutting a seam was at the miner's expense.

Mine Safety

Safety was always a concern of the miners but before 1900 not always of the owners. The most frequent cause of injury and death was roof falls. The roof generally consisted of black shale, which was soft and often fractured. Collapse was common in longwall mining if roofs were not propped up or if the props were accidentally knocked loose during mining or hauling. Another cause of accidents was runaway pit cars. Because of the undulating nature of many coal seams, floors rose and fell in roller-coaster fashion. If brakes on pit cars were not properly set or if couplings failed, miners and animals were at serious risk. Another cause of death was asphyxiation by toxic gases. These gases, collectively referred to as damp, consisted primarily of carbon dioxide, carbon monoxide, nitrogen, and water vapor, and they were generated by human and animal respiration, blasting, and engine exhaust. Ventilation, required to disperse toxic gases and bring in fresh air, was not always adequate. In 1901 twenty-nine miners died in mine accidents, and fifty-nine were injured (Davis 1990).

Unfortunately, in the early years of Iowa coal mining, many mine operators appeared to have little regard for the health and safety of miners, especially those who were foreign born. One corporate official, responding to concerns for the safety of miners, commented, "After all, it's not so serious, because most of the men killed are ignorant foreigners who can easily be replaced" (Schwieder 1983). Poor living conditions, low wages, and safety concerns led the miners to organize. Early attempts met with frustration because of the power of the owners. For example, in Appanoose County the Mystic Coal Company had been operating coal machines in its mines for several years. In 1909 a labor dispute arose over the machine wage scale for loaders. A mining arbitration board was called in, and it raised the scale by 10 cents. The owners retaliated by pulling the machines from the mines, stating that profit was too marginal. This decision worked

4.9. Cramped coal mining conditions. *State Historical Society of Iowa, Iowa City.*

a hardship on the miners. While following the machines they could make from $2.50 to $4.00 per day; having to mine with picks reduced their wages to $1.00 per day (Lees 1909). After numerous unsuccessful attempts at unionization, not only in Iowa but throughout the United States, the United Mine Workers of America was established in 1890, and living and working conditions gradually improved (Schwieder 1983).

Production

The years 1895 to 1925 were the most productive for coal mining in Iowa. In 1895 342 mines employing 6,363 miners operated in the state. Iowa ranked fifteenth in the nation in tonnage of coal produced. By 1917 Iowa had risen to fifth in coal production with nearly 9 million tons, while employing 18,000 miners (Drake and Ririe 1975). In 1925 the number of

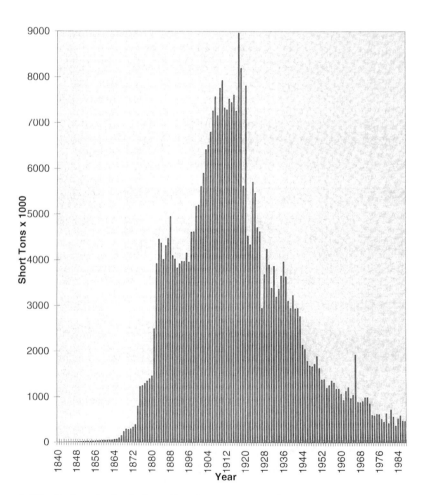

Coal Production in Iowa

4.10. Coal production in Iowa. *Data from the Geological Survey Bureau, Iowa Department of Natural Resources and the United States Bureau of Mines.*

128 *Iowa's Mineral Industries*

4.11. Coal strip mine. Mahaska County. *Geological Survey Bureau, Iowa Department of Natural Resources.*

mines (354) remained nearly the same as in 1895, but the number of miners had decreased since 1917 to about 11,000.

Coal production in Iowa began a serious decline in about 1930. The Great Depression reduced demand for coal nationwide. The railroads, which had been principal users of coal, began the switch to diesel and other fuels. Furthermore, most industries were established in locations that were distant from Iowa coal. Costs of transporting coal from Iowa to users became a serious issue. By the mid-1950s only a handful of mines remained in the state. The last pony mine (Appanoose County) closed in 1971 (Howes 1992) (fig. 4.10). In 1973 800,000 tons of coal were mined from five strip and two deep shaft mines (Drake and Ririe 1975). The last coal mine in Iowa ceased production in 1994.

With the advent of more powerful mechanized machinery and with the increased costs of underground mining, shaft and slope mines were gradually replaced by strip mines (fig. 4.11). The major conversion took place during the 1940s. Strip mining began with removal of overburden, the soil and rock that rested atop the coal. This was accomplished with a dragline or scraper. Coal was removed from the working face with a power shovel and loaded into trucks for haulage. The overburden was cast behind the working face, resulting in an ever-enlarging field of spoil piles.

Environmental Effects of Coal Mining

While strip mining afforded greater overall safety to miners, it created another kind of safety problem. Iowa coal contains a relatively high concentration of sulfur, a significant part of which is tied up as iron sulfide, primarily pyrite (FeS$_2$). The organic-rich shale that typically overlies Iowa coal is also pyrite-rich. Because of the method of overburden removal, which placed the first layer removed (the soil) at the bottom of the spoil pile and the last layer removed (the pyritic shale) at the top, pyrite was exposed to the atmosphere. Fine-grained pyrite quickly decomposes in the presence of water and oxygen. One of the products is sulfuric acid (H$_2$SO$_4$), an effective conveyor of toxic metals that gives rise to the problem of acid mine drainage. Rainwater leaching acid and dissolved metals from coal spoil piles carries these toxic materials into nearby streams and ponds, resulting in serious losses of plants and animals that depend on those waters. Lack of vegetation on spoil piles results in erosion, which exposes more pyritic shale to the atmosphere, perpetuating the problem.

Increasing awareness of the risks of surface coal mining to environmental safety in the late 1960s, and government support in the 1970s gradually led to changes in the nature of strip mining. Overburden was removed and set aside for replacement in the original stratigraphic order, thereby burying the pyritic shale and isolating it from the effects of atmospheric oxidation. In 1973, under the auspices of Iowa State University, the Energy and Mineral Resources Research Institute (EMRRI) was formed. Investigations performed at EMRRI and at the University of Iowa demonstrated that, with proper replacement and landscaping of overburden, coal-mined lands could be returned to agricultural and other uses (fig. 4.12).

Underground coal mining also presented environmental hazards. Ex-

4.12. Reclaimed strip mine land. Mahaska County. *Geological Survey Bureau, Iowa Department of Natural Resources.*

4.13. Sinkhole caused by the collapse of the land surface into an underground coal mine. What Cheer, Keokuk County, 1981. *Geological Survey Bureau, Iowa Department of Natural Resources.*

posure of pyritic material to the atmosphere generates acidic water which, when pumped to the surface, may contaminate surface waters. Lands over underground mines are subject to subsidence (fig. 4.13). It is estimated that as many as 6,000 underground mines operated in thirty-eight Iowa counties, potentially affecting about 32,000 hectares (Howes and Culp 1989). As many as 1,550 urban hectares alone are threatened by mine subsidence (Van Dorpe and Howes 1986). This hazard underscores the need to identify accurately the locations of Iowa's underground mines. The Iowa Abandoned Mine Lands (AML) and the Mineral Industry Location System (MILS) programs, administered by Iowa's Geological Survey Bureau, support studies of subsidence-prone areas. In the 1980s and the early 1990s most underground mines in Iowa were identified.

The Future

Strippable and deep reserves of coal still remain in the state (Landis and Van Eck 1965; Garvin and Van Eck 1978). Over 41 percent of Iowa is underlain by coal-bearing rocks, though much of this coal lies too deep to be commercially exploitable at present (Anderson 1979) (fig. 3.9). It is estimated that only 20 percent of the original strippable reserves of coal greater than 70 centimeters thick have been mined. Nationwide demand for coal is higher than ever before. Almost 60 percent of the country's electricity demand is met by coal. Given these facts, why has coal mining in Iowa diminished to the point of extinction? The answer lies in the environmental costs of mining (including reclamation) and burning coal. Iowa coal, though of relatively high BTU rank, is also high in sulfur and ash content and at present simply is not competitive with the cleaner-burning coals of the western United States. In the future, the value of coal as a source of organic chemicals may exceed its value as a fuel.

Gold

Although gold mining did not achieve the status of an industry in Iowa, its brief history is worth mentioning. The first reported discovery of gold in the state was made in 1853 by John Ellsworth, who claimed to have found gold on his farm along the Iowa River a short distance south of Eldora in Hardin County (fig. 4.14). Word of the discovery spread quickly and, not surprisingly, the deposit got richer and richer with the telling. The timing was right for gold fever in Iowa. The year 1853 was not a good one for farming in the Midwest, and memories of the famed California gold rush of 1848 were still fresh in people's minds. Some Iowa farmers left their fields with picks, shovels, and gold pans in hopes of striking it rich in Hardin County. As news of the discovery spread, prairie schooners began arriving with gold seekers from more distant parts. It is estimated that as many as 3,000 individuals played the gold lottery that year; all left disappointed. Frustration ran high. Local residents were accused of hiding the whereabouts of gold from out-of-town seekers, and John Ellsworth even received threats on his life (Petersen 1958).

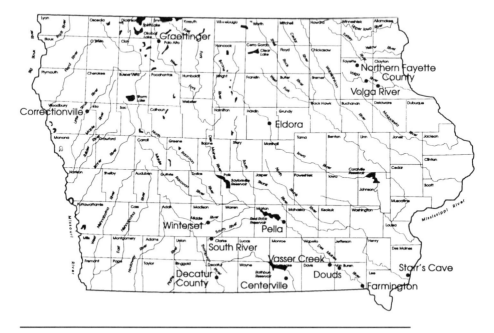

4.14. Some of the reported occurrences of gold in Iowa.

The fever soon abated, and farmers returned to Iowa's real gold—the soil. In 1857 when O. M. Holcomb, the first publisher of the *Hardin County Sentinel*, who had been prospecting along the Iowa River all summer for gold, reported a discovery, nobody got excited (Moir 1911).

In 1877 gold was again reported in Hardin County, this time about 11 kilometers north of Eldora. Glittering flakes were discovered among black magnetite sand along the Iowa River. A local jeweler pronounced it as "true grit." However, prospectors could only work out about 50 cents of the metal from each pan of black sand (Union Publishing Co. 1883; Moir 1911; *Des Moines Register* 1963; *Waterloo Sunday Courier* 1973). Again, Eldora turned out not to be El Dorado.

In 1881 a gold discovery was reported on Bear Creek near Ottumwa in Wapello County. Promoters soon organized the Wapello Gold Mining Company, which proceeded to set up a boiler and to identify a site for a proposed smelter. Another discovery was reported on Village Creek, and people began to flock to the "goldfields" to stake their claims. Enterprising businessmen, like O. J. Brisco, the principal promoter, were only too happy to accommodate them. Land prices soared. Some levelheaded Ottumwa businessmen thought it prudent to obtain an assay of the ground. The pay-dirt was reportedly sent to an assay office in Chicago, in the tender care of several prominent Ottumwa citizens. Gold fever continued to rise, as anxious would-be miners awaited the assay report. The report showed the "rich" dirt contained traces of silver and iron but no gold. Dreams of a bonanza evaporated—and so did O. J. Brisco (*Iowan* 1954).

Other discoveries of gold have been reported in Iowa. These include those at Centerville (Appanoose County); along Otter, Brush, Alexander, and Bear creeks (Fayette County); along the Volga River (Fayette and Clayton counties); at Starrs Cave (Des Moines County); along the Des Moines River south of Pella (Marion County); at Graettinger (Palo Alto County); at Douds and Farmington (Van Buren County); along Vasser Creek (Davis County); along the South River (Clarke County); at Winterset (Madison County); at Correctionville on the Little Sioux River (Woodbury County); and at several places in Decatur County (Keyes 1894b; Swisher 1945; *Centerville Iowegian* 1958; Petersen 1958; Anderson 1987) (fig. 4.14). Fowler Vina reportedly panned the equivalent of $18 in three days from the Volga River (Toney 1964). In the 1904 annual report of the Iowa Geologic Survey, Savage states that a "patient washer" could earn about $1.00 to $1.25 per day from the gravels of Otter Creek (Savage 1905).

It is likely that gold occurs along most of the major watercourses in Iowa, particularly in the eastern half of the state, because of the wide distribution of glacially transported granitic rock from which it was probably derived. However, there is little likelihood that gold will ever be found in sufficient concentration to be commercially exploitable.

Gypsum

The utilization of gypsum for plaster can be traced back at least to 12,000 B.C. in the Middle East, where it was employed principally for plastering walls and for mortar (Cody et al. 1996). The ancients learned that gypsum rock could be partially dehydrated by heating, then powdered and mixed with water to form a workable paste that hardened upon drying. Later it was learned that gypsum plaster has effective fire-retarding capability because of its low heat conductivity and because heating the plaster drives off water, which keeps the temperature below the ignition point of combustible material until dehydration is complete.

Gypsum plaster reportedly was introduced into the United States by Benjamin Franklin in 1785, who encountered plaster of Paris in France, where it was used as a wall finish, a casting material, and a soil nutrient. He demonstrated its value as a fertilizer by applying gypsum to a clover field in a pattern that produced the words "Land Plaster Used Here." The clover words were greener and more luxuriant than the surrounding plants. Gypsum mining and processing in the United States began in the state of New York in 1835. With westward expansion and new discoveries of gypsum resources, gypsum processing plants appeared in Michigan and Ohio (Wilder 1919).

Gypsum was first reported in Iowa in the Fort Dodge area by the pioneering explorer and mapper David Dale Owen, who considered the deposit the largest west of the Appalachians. Gypsum was identified during early geological studies of Iowa by A. H. Worthen (Worthen 1858) and C. A. White (White 1870), both of whom considered Iowa's gypsum to have great economic potential (fig. 4.15).

The gypsum industry in Iowa began at least by the 1860s with the quarrying of gypsum, which was used chiefly as a building stone. The softness of the stone made it easy to work, and it found its way into foundations and both interior and exterior walls (Cody et al. 1996). However, it was soon discovered that Iowa's high humidity and the relatively high solubility of gypsum in water caused exterior stones to deteriorate quickly. Limestone soon took its place.

Production of plaster in Iowa began in 1872. Gypsum rock was quarried along the Des Moines River and Two Mile Creek (then called Gypsum Creek), from bluffs where overburden was minimal. Development of the industry was facilitated by the arrival of the Illinois Central Railway at Fort Dodge. George S. Ringland, Stillman T. Meservey, and Webb Vincent established the Fort Dodge Plaster Mills, the first gypsum processing plant west of the Mississippi River (Rodenborn 1972). The plant, built at the head of Gypsum Creek, was designed for preparing gypsum for use as stucco. In 1873 the company name changed to the Cardiff Plaster Mills, the name taken from the Cardiff Giant (see chapter 5), and soon

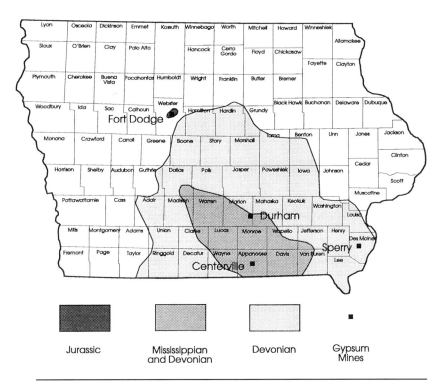

4.15. Bedrock areas of gypsum occurrence and areas of former and current gypsum mining in Iowa. *After McKay (1985).*

after the company was awarded a contract for 30,000 barrels of stucco for use in the new Illinois state capitol building. The partnership dissolved in 1882, and the Iowa Plaster Company (IPC) was formed, with Ringland, Meservey, and Vincent as chief officers. In 1882 IPC obtained a contract to supply all the plaster for the new Iowa state capitol building (Rodenborn 1972). In addition to its use as stucco, gypsum also proved valuable as a fertilizer. Because of its high solubility in water, gypsum readily breaks down in the soil to form sulfur, a valuable nutrient. It modifies soil pH, which helps optimize conditions for plant growth.

Extraction and Processing

In the early days of quarrying and processing gypsum, overburden was removed by horse-drawn scrapers (fig. 4.16). The exposed rock was then drilled, blasted, and hauled by team and wagon to the mill (fig. 4.17). At the mill the gypsum was crushed, dried if necessary to remove excess moisture, and ground to a fine powder. The heart of the gypsum mill was the kettle, consisting of a steel cylinder with an arched bottom, which was encased in brick. The earliest kettles could accommodate about 2 tons of gypsum. Heat was supplied to the bottom and sides of the kettle from a coal-stoked firebox. Gypsum powder was introduced into the kettle and heated to about 43°C, at which temperature water began to be released from the gypsum, causing it to "boil," a process called calcining. The temperature of the mass was maintained at about 104°C until most of the water was driven off. Stirring during the approximately two hours of calcining helped assure uniformity to the finished product. About one and a half of the original two molecules of water were released during calcining, producing bassanite ($2CaSO_4.H_2O$) or hemihydrate ($\beta CaSO_4.0.5H_2O$). Temperature control was important, for if the bottom of the kettle became too hot, the batch "burned" and was ruined. The calcined powder was tapped from the sides and bottom of the kettle, then cooled and packed in 107-kilogram wooden barrels for shipment (McKay 1992; Cody et al. 1996).

Finish plaster was not developed until 1884. Before this, the set time was too short, making the plaster difficult to work. George Ringland patented a process in which a chemical retardant was added to increase the set time, and the finish plaster industry was born. Other companies obtained patent licenses to use Ringland's process, a few without paying pat-

4.16. Eroded surface of gypsum beds. Fort Dodge, Webster County, 1917. *Calvin Photographic Collection, Department of Geology, University of Iowa.*

4.17. Early gypsum quarry. Fort Dodge, Webster County, 1902. *Calvin Photographic Collection, Department of Geology, University of Iowa.*

ent royalties. This led to five years of litigation, which eventually brought the pirates under control (Rodenborn 1972).

Production

From 1881 to 1891 annual production of raw gypsum in Iowa grew from 6,000 to 31,000 tons. By 1893, with 50,000 tons produced annually, Iowa ranked third in the production of raw gypsum after New York and Michigan (McKay 1992). As quarrying moved greater and greater distances away from valley walls, increased costs of quarrying due to greater thickness of overburden led to a shift to underground mining. The first mine in Iowa was opened by the Cardiff Gypsum Plaster Company in 1895. A vertical shaft was driven to the base of the gypsum bed, and mining was by room and pillar. Water was channeled to sumps and then lifted to the surface by pumps, and workings were ventilated by large electric-powered fans. Quarries closer to river bluffs were converted to drift mines. At one operation, where the mine and the mill were on opposite sides of the Des Moines River, gypsum rock crossed the river by an overhead cable tram from mines on the west side to the mill on the east; at another operation a pontoon bridge was used (Wilder 1919). Between 1895 and 1901 three new gypsum mines were opened—all under ground.

Growth in the industry brought another change. Early independent mines and processing plants consolidated to form combines or were bought out by eastern companies. In 1882 the first combine, the Iowa Plaster Association, formed from the Iowa Plaster Company and the Fort Dodge Gypsum Stucco Company (Rodenborn 1972). In 1902 the United States Gypsum Company (USG) was formed through the consolidation of some thirty firms nationwide, including four of the six in Iowa (McKay 1992). USG eventually acquired six Iowa-originated companies (Rodenborn 1972) (table 4.1).

In 1907 the Sackett Plasterboard Company began production of plasterboard, the forerunner of wallboard (drywall). This product consisted of seven layers of paper alternating with six cores of gypsum, the whole pressed into a sandwich 0.8 centimeter thick. Plasterboard was designed

Table 4.1. History of Gypsum Product Operations at Fort Dodge, Iowa

Company	Place of Origin	First Year in Iowa	Method of Extraction	Status
Fort Dodge Plaster Mills (Cardiff Plaster Mills)	Fort Dodge, Iowa	1872	quarry	acquired by Iowa Plaster Co. 1882
Iowa Plaster Co.	Fort Dodge, Iowa	1882	quarry	combined to form Iowa Plaster Assoc. 1891
Fort Dodge Gypsum Stucco Co.	Fort Dodge, Iowa	1884	quarry	acquired by Iowa Plaster Assoc. 1891
Iowa Gypsum Manufacturing Co. (Duncome Stucco Mills)	Fort Dodge, Iowa	1889	quarry	acquired by U.S. Gypsum Co. 1902
Iowa Plaster Assoc.	Fort Dodge, Iowa	1891	quarry	acquired by U.S. Gypsum Co. 1902
Cardiff Gypsum Plaster Co.	Fort Dodge, Iowa	1895	shaft mine	company dissolved 1961; interests bought by Flintkote Co.
Fort Dodge Plaster Co.	Fort Dodge, Iowa	1899	shaft mine	acquired by U.S. Gypsum Co. 1902

Continued

Table 4.1. *Continued*

Company	Place of Origin	First Year in Iowa	Method of Extraction	Status
Mineral City Plaster Co.	Fort Dodge, Iowa	1899	shaft mine	acquired by U.S. Gypsum Co. 1902
Carbon Plaster Co.	Fort Dodge, Iowa	1900	shaft mine	acquired by U.S. Gypsum Co. 1902
U.S. Gypsum Co.	New Jersey	1902	quarry and shaft mine	currently operating (1997)
Plymouth Gypsum Co.	Fort Dodge, Iowa	1903	drift mine	acquired by Universal Gypsum and Lime Co. 1923
Hawkeye Plaster Co.	Fort Dodge, Iowa	1906	never mined	acquired by American Cement Plaster Co. 1909
American Independent Gypsum Co.	Fort Dodge, Iowa	1906	shaft mine (?)	acquired by Acme Cement Plaster Co. 1910
Iowa Hard Plaster Co.	Fort Dodge, Iowa	1906	shaft mine	acquired by American Cement Plaster Co. 1910
Sackett Plaster Board Co.	New York	1907	plant only	acquired by U.S. Gypsum Co. 1913
Acme Cement Plaster Co.	St. Louis	1908	mill only	acquired by Certain-teed Products Corp. 1928
American Cement Plaster Co.	Lawrence, Kans.	1909	shaft mine	acquired by Beaver Products Co. 1922
Wasem Plaster Co.	Fort Dodge, Iowa	1909	shaft mine	acquired by Celotex Corp. 1950
Iowana Gypsum Co.	Fort Dodge, Iowa	1920	mill only	acquired by Universal Gypsum & Lime Co. 1923
Beaver Products Co.	Tonawanda, N.Y.	1922	wallboard plant only	acquired by Certain-teed Products Corp. 1928
Universal Gypsum & Lime Co.	Delaware	1923	shaft mine, quarry	merged with National Gypsum Co. 1935
Certain-teed Products Corp.	Maryland	1923	shaft mine, quarry	acquired by Bestwall Gypsum Co. 1956
National Gypsum Co.	Delaware	1935	shaft mine, quarry	currently operating (1997)
Celotex Corp.	Delaware	1950	room and pillar mine	currently operating (1997)
Bestwall Gypsum Co.	Chicago	1956	acquired existing holdings	merged with Georgia Pacific Corp. 1965
Flintkote Co.	New Jersey	1960	shaft mine	unknown
Georgia Pacific Corp.	Georgia	1965	acquired existing holdings	currently operating (1997)

Sources: Wilder (1902) and Rodenborn (1972).

Table 4.2. Fire Damage to Fort Dodge Gypsum Operations

Year	Company	Results of Fire
1909	United States Gypsum Mineral City Mill	damage to wood fiber bin
1912	Acme Cement Plaster Co.	mill destroyed
1913	American Cement Plaster Co.	mill destroyed
1918	Wasem Plaster Co.	mill, blacksmith shop, and wood fiber building destroyed
1937	National Gypsum Co.	mill and warehouse destroyed

Source: Rodenborn (1972).

to replace plaster lath and rough coat. It had the advantage of ease of installation and greater fire resistance.

In 1910 gypsum was discovered near Centerville in Appanoose County during coal exploration drilling by the Scandinavian Coal Company. Further drilling demonstrated that commercial quantities of gypsum were contained in Mississippian strata at a depth of about 165 meters. A shaft was completed in 1912, but because of problems with water, further work was not done until 1917. The Centerville Gypsum Company installed a mill that operated intermittently until 1922, when the company sold out to USG, which continued operation of the mine and mill until 1934. The total production was small (Wilder 1919; McKay 1992).

By 1925 annual production of Iowa raw gypsum had reached 800,000 tons with a value of $6.7 million, which accounted for 29 percent of Iowa's nonfuel mineral production. Uses of gypsum products included various plasterboard and wallboard designs; fireproof gypsum blocks; integrated systems for assembly of gypsum-based floors, ceilings, and roofs; and as a retarder of set time in concrete (McKay 1992). Gypsum companies in Iowa prospered in spite of markets that were at times sluggish and in spite of operating expenses that not infrequently included losses due to fires. Mill fires were common because of the intense heat generated during the calcining and plasterboard drying processes and due to an abundance of paper and combustible wood fiber, which were stored in the mills (table 4.2). Gradually, wood in mills was replaced by steel and concrete, and fire susceptibility was greatly reduced (Rodenborn 1972).

With the advent of mechanized equipment capable of moving large quantities of unconsolidated material and rock per unit of time, gypsum mining gradually returned to the surface. USG led the way in 1927; National Gypsum opened its first quarry in 1944. As underground operations came to a close, pillars were shaved or altogether removed, a process referred to as robbing. Loss of mine roof support resulted in the gradual collapse of mine workings and the consequent formation of surface craters. Although most areas of collapse were subsequently removed during strip mining, mining-induced sinkholes currently can be seen south of Fort Dodge (Cody et al. 1996).

The Great Depression had a profound effect on the gypsum industry nationwide. By 1933 production in Iowa had declined to $1.36 million. It would not recover to pre-Depression levels until the close of World War II (McKay 1992). The economic boom following World War II stimulated further exploration for gypsum. In 1957 gypsum was discovered during water well drilling near Mediapolis in east-central Iowa. A shaft was sunk to a depth of 185 meters, and the Sperry Mine began production on June 20, 1960 (fig. 4.18). The gypsum at Sperry occurs in middle Devonian limestones. By 1990 production at Sperry reached 2.19 million tons at a value of $14.2 million, or about 4.5 percent of the mineral production in Iowa (McKay 1992) (fig. 4.19).

At present four companies mine and/or mill gypsum in the Fort Dodge area: USG, National Gypsum Co., Georgia Pacific Corp., and Celotex Corp. (table 4.1). Gypsum reserves at Fort Dodge are sufficient for about thirty years at current rates of consumption (Cody et al. 1996). USG is the only company operating at Sperry. Reserves at Sperry are rated at a hundred years (McKay 1992). The Kaser Corporation currently produces gypsum of Mississippian age from the Durham Mine in Marion County (McKay 1985). The presence of gypsum in numerous drill cores and cuttings in both Mississippian and Devonian rocks in a large area of the state indicates that the potential reserves of gypsum in Iowa are large.

Further information on the history of the gypsum industry in Fort

4.18. Loading gypsum for hauling from the Sperry Mine. Sperry, Des Moines County. *United States Gypsum Company.*

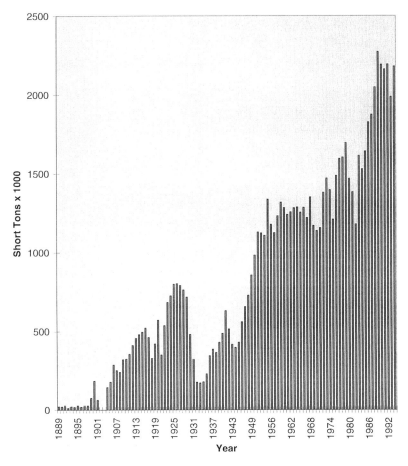

4.19. Gypsum production in Iowa. *Data from the Geological Survey Bureau, Iowa Department of Natural Resources and the United States Bureau of Mines.*

Dodge can be found in Frank Wilder's report to the Iowa Geological Survey (Wilder 1919), in a book by Leo Rodenborn written in commemoration of the hundredth anniversary of the plaster industry in Iowa (Rodenborn 1972), and in Cody et al. (1996).

Iron

The presence of iron ore in Iowa was noted during early surveys of the state. Hall and Whitney (1858) describe scattered occurrences of limonite and hematite in the coal measures of southern Iowa and along streams in Jackson County. None of the occurrences were considered to hold any promise of commercial development.

The date of discovery of iron ore deposits at Iron Hill, near Waukon in

Allamakee County, is not known, but the beginnings of mining can be traced back to 1882 (Rehder and Cook 1972) (fig. 3.13). Under the super-intendency of a Mr. Coe, an initial trench measuring 3 x 1 x 2.5 meters was dug with pick and shovel. By December 13 thirty shafts were sunk, the two deepest to about 10 meters. The ore, judged to be 7 to 10 meters thick, was to be hauled to the surface by rope and windlass. Though the ore was of comparatively low grade, its proximity to the Mississippi River and the general downward incline from Waukon to the river convinced promoters that it could be mined profitably. Early assessments of the size of the ore body indicated a hundred-year reserve. Laborers were hired at $1.50 per day, and Coe's company paid out about $2,500 in exploration expenses. Operations ceased on December 27, 1882, due to cold weather. Coe did not return in the spring of 1883 (Rehder and Cook 1972).

By 1895 the Waukon mines had not yet produced commercial ore. The principal problem appeared to be lack of suitable means of transportation from the mines to potential users. In 1898 George S. Finney dug numerous test pits and submitted samples for analysis. Having obtained favorable assays, in May 1899 Finney obtained a twenty-year lease on the mining property. A second party was to pay 10 cents per ton royalty on all ore mined, with 10,000 tons as the minimum annual output. Commercial pro-duction began in 1899 with an output of 1,260 long tons of ore, which was shipped to furnaces in Milwaukee (Rehder and Cook 1972).

The Waukon Iron Company was organized in April 1900 with a capital stock of $50,000, which increased to $500,000 by June of that year. Since there was little overburden, the ore could be worked by open pit methods. Rock was reduced by black powder and dynamite (for hard boulders). The ore was then loaded into two wooden ore cars and hauled by horses to an on-site crushing and washing plant. Later, ore was hauled out of the mine on a rail spur, one car at a time. Each car was shoved upgrade by the engine to the plant, unloaded, and then lowered downgrade by cable back to the mine. Water for washing was obtained from a 150-meter well and pumped into a 12,000 barrel–capacity storage reservoir. The washed ore was carried by pan conveyor to storage bins. Chert nodules were removed by hand during transport (Rehder and Cook 1972).

The ore was easy to mine but expensive to ship. The cost of shipping the ore the 5-kilometer distance from the washer to the nearest railroad was 50 cents per ton, which was more than it cost to ship much richer Lake Superior iron ore from Duluth to Cleveland. The mining operation faced another problem. Waste water and tailings from the washing plant emptied into a ravine and caused siltation and increased flooding of nearby farm fields and water pollution. Angry farmers went to court and obtained an injunction against the Waukon Iron Company. The company ceased opera-tions in 1902 (Rehder and Cook 1972).

In that same year William Ingram appeared on the scene. A promoter

4.20. Iron washing plant. Waukon Junction, Allamakee County, ca. 1904. *Calvin Photographic Collection, Department of Geology, University of Iowa.*

of interurban railroads, he incorporated the Iowa Hematite Railway Company (IHRC), the primary purpose of which was to get ore from Waukon to the Mississippi River. Investment capital was set at $25,000, but soon increased to $250,000, with authorization to increase it to $1,500,000. Franchises were obtained from the towns and the county. That was last anyone heard of Ingram and the IHRC (Rehder and Cook 1972). It appeared that Iowa was destined not to ship its iron.

About 1904 or 1905 the Missouri Iron Company (MIC), with headquarters in St. Louis, began exploration and development at Iron Hill. Development included the installation of a modern washer with the capacity to produce 600 tons of finished ore per day (fig. 4.20). The ore was to be washed, jigged, and roasted to improve the iron concentration to 55 to 60 percent. With E. F. Goltra as MIC president and R. W. Erwin as general manager, the company obtained a permit to operate in 1907. Because of the previously mentioned problems with waste water and tailings disposal, MIC decided to build the wash plant at Waukon Junction, over 40 kilometers east, where disposal would be less disruptive to other land use. Ore was to be freighted by team and wagon from Waukon to Waukon Junction. The wash plant was installed on the side of the bluff overlooking Waukon

Junction. Soon MIC decided to build a railroad spur up the bluff to the wash plant, instead of hauling the ore by inclined tramway. About 70 meters of grade was blasted out of the bluff, when company officials changed their minds again. The development of a new dry method of ore concentration obviated the need for a wash plant with its attendant waste disposal problems. Moving the plant back to Waukon would eliminate the need to haul both ore and waste (about half the total tonnage) to Waukon Junction. Leaving a scarred landscape in its wake, MIC moved the operation to Waukon (Rehder and Cook 1972).

In 1910 a rail line was installed from the new plant site to the Waukon Branch of the Milwaukee Railroad. The line was built with team and wagon at a cost of about $100,000 (for a length of 8 kilometers). Completion of the railroad allowed heavy machinery to be brought to the mine, including a 40-meter rotary kiln. Buildings at the plant included a drying and crushing house, roaster, air washer, reducer, magnetic separator, loading bins, office, laboratory, refuse bins, coal storing and drying house, power house, and machine shop (Rehder and Cook 1972).

Since the overburden was shallow, stripping was not needed, and all rock was sent to the crusher. Rock was loosened by drilling holes at about 25-centimeter centers, which were loaded with dynamite that was detonated by batteries. Jackhammers broke up the large boulders. Rock was loaded onto electric-powered dump cars. Ore was concentrated by the Goltra process. In this process, the ore was dumped on 30-centimeter grids, with oversized boulders reduced by sledges. The ore was then dried to remove free water and passed over a 6-centimeter grid; the oversized material was sent through a crusher. Next, the ore passed along a conveyor system to the roaster, where chert and other impurities were removed en route. Operating at a temperature of 870 to 980°C, the roaster drove off structural water, thus converting limonite to hematite. The hematite (Fe_2O_3), while still hot, was sprayed with crude oil under pressure, which created a chemically reducing environment and transformed the hematite into magnetite (martite [Fe_3O_4]). The reduced ore was then cooled and passed through a magnetic separator to remove any remaining nonmagnetic waste material. The recovery averaged about 50 percent, with a minimum concentration of metallic iron of 55 percent (Howell 1915).

Production began, at last, in 1913 with an expected output of 450 tons per day. Concentrates were shipped to Milwaukee, Chicago, and St. Louis. Unfortunately, about this time the iron industry was experiencing a depression, and production at MIC soon ceased. The plant reopened in 1915 with production of 350 tons per day (MIC's most productive period). In 1916 MIC changed its name to the Mississippi Valley Iron Company (MVIC), with E. F. Goltra as president (Rehder and Cook 1972).

In 1917 the United States government took control of the railroads. This move caused serious and permanent damage to the Waukon iron industry.

Rail cars were allocated on a perceived need basis. The MVIC found itself in competition with the Lake Superior iron district in northern Minnesota. The ores of the Mesabi Range were richer than those at Waukon; thus, they were in high demand. Soon, MVIC could not obtain cars to haul its ores. Operations declined in 1918 but continued part-time until 1922, when production permanently ground to a halt. Minimum security was maintained for a few years, but the buildings and machinery gradually disintegrated. In 1934–1935 the railroad was dismantled. The MVIC went into receivership, and the 6,000 tons of treated ore remaining at the plant were auctioned off in 1941 (Rehder and Cook 1972).

World War II brought renewed interest in mineral resources throughout the country, including the upper Midwest. The U.S. Bureau of Mines (USBM) did exploratory work for lead, zinc, and iron in northeast Iowa. Despite rather large reserve estimates (original crude ore reserves were estimated at 10 to 12 million tons, most of which was still in the ground), the USBM determined that Waukon iron was too low grade to be competitive with other sources in the upper Mississippi Valley and Great Lakes regions. In 1950 a steel mill in Gary, Indiana, bought some of the stockpile, and in 1958 the rest was purchased by the Marquette Cement Company of Des Moines, Iowa (Rehder and Cook 1972). Iron ore remains buried under soil and vegetation at Iron Hill awaiting the time when market conditions may again favor its exploitation.

For a brief period iron was mined at another locality in Iowa. During the lead and zinc mining operations in the Dubuque area, local concentrations of limonite (ocher) were discovered. These deposits resulted from the vadose weathering of preexisting iron sulfide minerals. At the Larkum range near Durango, abundant ocher was found in an east-west crevice about 360 meters long. The ore was considered exploitable for use in mineral paint and as a possible iron source (Calvin and Bain 1899). A limonite ore company was formed to exploit the mineral, and in 1896–1897 250 tons of ore were shipped from Durango. However, due to its low grade it was not competitive with other sources, and operations soon ceased.

Lead and Zinc

The best summary of the history of lead and zinc mining in Iowa probably is that written by Charles Reuben Keyes in his "Historical Sketch of Mining in Iowa" (Keyes 1913; see also Langworthy 1854–1855; Leonard 1896; Shiras 1902; Van der Zee 1915; Goodwin 1919; Petersen 1931). The time of the beginning of lead mining in Iowa depends in part on how one defines mining. According to Henry Schoolcraft, who was chief narrator for an exploring expedition to the upper Midwest in the early nineteenth century, lead was first discovered in 1788 by the wife of Peosta, a Meskwaki (Fox) warrior of Kettle Chief's band. While it is the popular belief that French entrepreneur Julien Dubuque was the first lead miner, there is ample evidence that the discovery and exploitation of lead occurred much earlier, indeed long before the arrival of Europeans. The availability at the surface of loose masses of galena, which are easily cleaved to reveal their bright, reflective interiors, suggests that the mineral likely was known to the ancient inhabitants of the upper Mississippi Valley. That early Native Americans were attracted to the lustrous galena is attested to by its frequent occurrence in archaeological remains of the Mound Culture (Grant 1979). The earliest lead miners clearly were not Europeans.

Early French Influence

The westward expansion of the French fur industry, coupled with the generally good relations between the French and the Native American tribes, brought French voyageurs and fur hunters into the upper Mississippi Valley. Fur hunting often required weapons and lead ammunition; thus, the meeting of Native Americans, the French, and lead was inevitable. The first recorded visit of a European to the lead region was made by Jean Nicolet, European discoverer of Lake Michigan, in 1634. Nicolet entered Green Bay, Wisconsin, and traveled down the Wisconsin River to its confluence with the Mississippi at the site of Prairie du Chien. Continuing his travels down the Mississippi, he passed through the lead region. The nature of his contact with the Sac and Meskwaki is not known with certainty, but it is hard to believe that he would not have learned of their lead mining activities. Later seventeenth-century French explorers, including Nicholas Perrot, reported the occurrence of lead and lead mining. Despite the common belief that Dubuque was first, Perrot is considered to be the first European to mine lead in Iowa (Van der Zee 1915). From these and other reports it appears that French mining and primitive smelting of lead were occurring at least by 1650. Mines and furnaces for processing ore were well established on both sides of the Mississippi and along its eastern tributaries by the end of the century. Sieur Pierre le Sueur secured a commission from Louis XIV of France to open additional mines, and he returned to the lead region with numerous miners in 1699. In the summer of 1701 he shipped 4,000 pounds of ore down the Mississippi and back to France (Keyes 1913). The Meskwaki reportedly were taught by the French how to work the deeper mines. Using a long-established method, they built fires to heat the rock and then shattered it with a dousing of cold water. After lifting the ore to the surface, they smelted the lead by placing the ore among logs, which were set afire. The heat was sufficient to produce some molten lead, which ran into a pit, where it solidified into large slabs (plats) weighing from 30 to 70 pounds each. Hundreds of tons of lead were produced by these crude methods (Petersen 1931).

French mining in the lead region continued throughout the first half of the eighteenth century. The French also smelted ore with essentially the same process used by the Meskwaki. The crude lead was remelted into bars of 60 to 80 pounds each. The mines were said to yield over 2,000 bars a year, during four or five months of work (Petersen 1931). France's fortunes in the upper Mississippi Valley changed dramatically after its defeat in the Seven Years War (French and Indian War) with England in 1763. As a result, Canada and all French-owned territory east of the Mississippi River were ceded to England. France secretly ceded all its territory west of the Mississippi to Spain. French traders were gradually displaced by English traders, and Spain controlled land grants west of the Mississippi.

The next actor in the lead mining drama was Julien Dubuque. Born in Canada in 1762, Dubuque came to Iowa from Fort Michilimackinac in northern Michigan and built a house near the mouth of Catfish Creek in the vicinity of the modern city that bears his name. He had a remarkable way with Native Americans, who called him Little Cloud, and in 1788 he was granted exclusive permission from the Meskwaki to mine lead on 60,000 hectares of their lands. This concession is said to be the "first conveyance of Iowa soil to the whites by the Indians" (Goodwin 1919). The written document (in French) specified that Dubuque was permitted to "work at the mine as long as he shall please . . . without anyone hurting him, or doing him any prejudice in his labors." The grant was confirmed in 1796 by Carondelet, governor of Spanish-held Louisiana, under the title "Mines of Spain." His claim was defined as follows: "from the margin of the waters of the little river Maquanquitois [Little Maquoketa] to the margin of the Mesquabysnonques [Tetes des Mortes], which forms about seven leagues [about 34 kilometers] on the west bank of the Mississippi, by three leagues [about 15 kilometers] in depth" (Shiras 1902) (fig. 4.21). He soon expanded his empire to include all the lead mining lands and their operations on the east side of the river in Wisconsin and Illinois. Mining was done with American Indian labor, but Dubuque brought in ten Canadians from Prairie du Chien to help him manage his operations (Goodwin 1919).

In 1802 France, having recovered from the Seven Years War, persuaded Charles IV of Spain to trade back the Louisiana Territories. In a political move calculated in part to lessen British influence in the New World, France sold Louisiana to the United States in the following year. The U.S. government soon commissioned the exploration of the newly acquired lands. In the late summer of 1805 Lieutenant Zebulon M. Pike, for whom two Pikes Peaks were named (one in northeast Iowa, the other in Colorado), traveled up the Mississippi and visited Dubuque's Mines of Spain. Pike was received cordially, but Dubuque was evasive about his mining activities and grossly understated his lead production. Pike left without actually visiting the mines and with little information about Dubuque's operations (Shiras 1902).

In 1807 the federal government passed a law declaring all mineral lands reserved for future mining. The law further stated that lands could not be

4.21. The lands of Julien Dubuque in the late eighteenth century. The original text of the map is in French. An English translation was added later. *From Shiras (1902).*

settled until they were surveyed and sold. Parcels of land were available for leases of three to five years, with royalties of 10 percent to be paid upon the value of minerals extracted. The 1807 law was largely ignored, especially where mining lands were concerned (Keyes 1917).

Dubuque's financial empire collapsed shortly before his death in 1810. He had sold lead to buy trade goods for the Meskwaki and other Native Americans. According to Van der Zee (1915), some believe that he was overly generous with them and as a result incurred insurmountable debts. In an effort to improve his financial condition, in 1805 he sold the southern half of his tract to one of his creditors, Auguste Choteau, for the reported sum of $10,848.60 (Van der Zee 1915). His creditors were unable to maintain the good relations he had developed with his Native American hosts, and soon after his death the Sac and Meskwaki burned his buildings, tore down his fences, and otherwise removed all traces of their former tenant's existence (Ludvigson and Dockal 1984). Meskwaki villagers took up mining, and they produced 400,000 pounds of lead in 1811 alone (Van der Zee 1915).

U.S. Mining Operations

American mining interests in the upper Mississippi Valley were stifled by the War of 1812, when the mines were controlled by the British and their Native American confederates. After the close of the war, the United States quickly built several forts along the Mississippi in an effort to secure a foothold. By 1822 American companies had begun mining operations on the east side of Mississippi River in what became known as the Upper Mines (to distinguish them from the Lower Mines of Missouri, which had been in production for several years). On the west side of the river, the Meskwaki continued working Dubuque's old claims. Whites were not permitted to work the mines until 1833, but the Meskwaki sold lead and lead ash to American traders (Hall and Whitney 1858).

By 1830 American mining companies were well established in Wisconsin and Illinois but not in Iowa because Iowa lands were still claimed by the Sac and Meskwaki. In 1830, because of hostilities with the Sioux, the Meskwaki abandoned the mines and sought protection near Rock Island, Illinois. Once their absence was discovered, miners from Illinois and Wisconsin slipped across the river and "appropriated" the mining operations. James and Lucius Langworthy were among the first to work the Mines of Spain. Competition for claims was so keen that the miners organized and drafted rules of governance. The federal government considered the miners trespassers on what was still Native American land, and it sent federal troops to drive the miners out. Their buildings were burned, and an army attachment was permanently assigned to the area in order to discourage future squatters. A short time later, under protection by U.S. army troops, the Meskwaki returned to their lands and took up mining again. They mined newly discovered deposits, reportedly producing a million pounds of ore from a single lode (Van der Zee 1915). Given the wanton seizing of Native American lands that did and would accompany the settlement of the United States, this act of protecting Native American rights seems out of character for the United States government. The protection, unfortunately, would be short-lived (Van der Zee 1915).

In 1830 the United States proposed to purchase the Mines of Spain from the Meskwaki, but the asking price was too steep. However, in 1831 the Meskwaki were again at war with the Sioux, and again they left the mines.

In early 1832 George Davenport appeared in Washington, D.C., with a proposal from the Sac and Meskwaki to sell the lead mines. Unfortunately, the government acted too slowly (Cole 1938).

Eighteen thirty-two was the year of the infamous and tragic Black Hawk War. In a futile effort to escape eviction from tribal lands, Chief Black Hawk and his people led U.S. army troops on a chase throughout western Illinois and southern Wisconsin. In the end, the Sac and Meskwaki were decimated by starvation and senseless killing, culminating at the Battle of Bad Axe on August 2, 1832. The treaty ending the fighting, which came to be known as the Black Hawk Purchase Treaty, was signed on September 21, 1832. It forced the Sac and Meskwaki to cede trans-Mississippian lands to the United States. These lands included Dubuque's Mines of Spain. In return for their lands the Sac and Meskwaki received $20,000 a year for the next thirty years. The treaty was not to take effect until June 1, 1833, but white settlers and miners refused to wait for the Meskwaki to leave. In September 1832 150 miners and their families invaded "the Savage Lands" (Cole 1938), and again the federal government intervened. This time Colonel Zachary Taylor (later to become the twelfth president of the United States) from Fort Crawford, Wisconsin, sent Lieutenant Jefferson Davis (later to head the southern Confederacy) and federal troops across the river to drive the squatters out (Cole 1938). Both of these men had been active participants in the Black Hawk War. Invaders, including some sent by Auguste Choteau from St. Louis, were driven out "at the point of the bayonette" several times during early 1833 (Van der Zee 1915). Early in the morning of June 1, white settlers and miners scrambled across the Mississippi to stake their claims. They were still in violation of the 1807 law requiring that lands be surveyed before they were settled, but the law was not enforced. The question of ownership of the Mines of Spain was complicated by the fact that Julien Dubuque and Auguste Choteau had a signed document naming the latter as the grantee upon Dubuque's death. French claimants, operating from St. Louis, demanded title to the lands. The United States government refused to acknowledge Choteau's document, stating that the original grant was merely a permission that pertained to Dubuque as an individual and that he did not have legal right to transfer title to the lands. The matter went to court, where the battle was fought for over forty years, up to the Supreme Court. In the end (1853) the government won.

Mining Methods

As previously stated, prior to Euroamerican settlement mining was done for many years by the Meskwaki using methods and tools acquired from the French. According to Schoolcraft (quoted in Calvin and Bain 1900), most mining was done by women and old men using hoes, shovels, picks, and crowbars. Ore was dug from shallow trenches and from short drifts dug into the hillsides. Langworthy (1854–1855) describes the mining process: "[The Meskwaki] would dig down a square hole, covering the entire width of the mine leaving one side not perpendicular, but at an angle of about forty-five degrees, then with deer skin sacks attached to a bark rope, they would haul out along the inclining side of the shaft, the rock and ore. Their mode of smelting was by digging into a bank slightly, then put up flat rocks in a funnel shape, and place the ore within, mixed with wood; this all burn together, and the lead would trickle down with small excavation in the earth, of any shape they desired, and slowly cool and become fit for exportation." Women carried the lead in baskets to the Mississippi River, where it was ferried across the river to furnaces on the east side and sold at the rate of $2.00 per 120 pounds. Lead ash was sold at $1.00 a basket.

Prior to the 1850s American miners extracted lead, as had their Native American predecessors, from shallow trenches and short drifts into hillsides. The earliest reports of mining methods come from David Dale Owen's pioneering survey of the upper Midwest (Owen 1844). By the mid-1850s small shaft mines appeared. These operations involved a windlass and a bucket for bringing up ore and for transporting miners to and from the workings (fig. 4.22). The lead ore was contained primarily in east-west–trending crevices. Shafts were driven to intersect them, and drifts followed them for as much as several hundred meters. In the Dubuque area the Langworthy Mine produced lead from a crevice over 1 kilometer long (though it was not productive over its entire length). Langworthy (1854–1855) describes the discovery of the deposit (which occurred prior to 1836): "The Langworthy lode exhibited the most astonishing specimens of lead ore ever found, perhaps, in any country. As the work-men, with the owners, penetrated into this cave for the first time, a hollow sound issued from it and the air came freshly from the west, in the direction of the vein. . . . They began to examine the cave for minerals, each man carrying a lamp or candle. Passing along through various windings and narrow spaces they suddenly came to one of the subterranean vaults which was completely filled with the shining ore, lighted up and sparkling like diamonds, or lying in great masses or adhering to the sides and roof of the cave in huge cubes."

Many of these crevices were unusually rich in galena. One near Hazel Green, Wisconsin, exposed a solid mass of galena 30 meters square. When reported by Whitney in 1857 (Hall and Whitney 1858), 600 tons of galena had been removed, and much more remained. At Dubuque, Levins Cave, named after its discoverer, was a natural opening 40 meters long, 6 meters high, and 6 to 9 meters wide. The entire ceiling appeared to have been originally encrusted with galena more than half a meter thick. One mass, which had fallen to the floor, was estimated to weigh 23,000 pounds. Other

4.22. Lithograph of mid-nineteenth-century lead mine shaft. *From Owen (1844).*
State Historical Society of Iowa, Iowa City.

masses of lead were found buried in the soft clay floor of the cave as much as half a meter below the surface. About 1,800 tons of lead were removed from this cave in three to four years (Hall and Whitney 1858). Langworthy (1854–1855) stated that 5,000 tons of lead ore were taken from his mine.

Lead was not the only metal of commercial interest in the Dubuque area; zinc was also abundant. During the early years of mining at Dubuque it was mined principally as zinc carbonate (smithsonite), which was commonly known as dry bone because of its color and porous earthy texture. As mining progressed to deeper levels, smithsonite gradually gave way to sphalerite, known to the miners as jack or blende, which upon oxidation yields dry bone. Large quantities of dry bone were mined from the Julien Avenue crevices (Calvin and Bain 1900). Early workings were also located at Durango, Sherrills Mound, and Guttenberg (fig. 4.23). At Guttenberg the ores were bedding-plane controlled, with a productive zone as much as 100 meters wide and 500 meters long. This ore was mined by room and pillar.

4.23. Goldthorpe's zinc mine. Durango, Dubuque County, ca. 1870. *Calvin Photographic Collection, Department of Geology, University of Iowa.*

As shafts went deeper the mines encountered water. Early pumping was done with a chain pump worked by hand and later by oxen lifting whiskey barrels full of water (Calvin and Bain 1900). In 1857 the steam-powered pump was introduced, which could raise several hundred liters of water per minute. The pump at the Karrick Mine was the only one on the west side of the Mississippi River used to drain a mine. It consisted of five wood-fired steamboat boilers which kept ten wagons hauling wood at the considerable cost of $100 per day (Calvin and Bain 1900). An undesirable side effect of the pumping was that it frequently caused nearby springs and shallow wells to go dry. By 1900 hand-operated mining equipment was generally replaced by power hoists, air presses and power drills, dynamos (the current was used to fire explosives), and electric lights.

Ore Processing

In the early days concentrating lead ore was a relatively simple task because the ore was usually free of wall rock and other impurities. Washing to remove surface coatings of iron oxide generally was sufficient to prepare the ore for smelting. Dry bone was hand sorted, washed with log washers, and hand picked. Blende (sphalerite) was crushed, heated, and jigged to remove attached marcasite and wallrock.

Lead-smelting operations go back to the time of Julien Dubuque. Dubuque erected several furnaces on his property. Each consisted of a stone platform on which alternating layers of wood and ore were stacked. The ores were roasted, which oxidized the sulfur and produced lead metal. Unfortunately, much of the lead was lost to oxidation (or lead ash). After Dubuque's death, the Sac and Meskwaki continued to smelt lead by this process. From these early beginnings smelting gradually became more efficient. Efficiency improved markedly with the arrival of the Scotch hearth, which was introduced into the United States in 1835. The second of these was built in Dubuque. Calvin and Bain (1900) describe the hearth as consisting of a well 35 x 50 x 65 centimeters, which was lined with cast iron. The ore and fuel (wood) were fed in alternately, and melting was promoted by a blast from the rear of the furnace. Half a cord of wood was required to reduce 300 pounds of ore. The metal ran down an apron into molds, which produced "pigs" of about 72 pounds each. A second furnace, fired by coke, reprocessed the slag from the first. A 72 percent recovery rate from the first furnace was improved to 90 percent with the second furnace. In some furnaces the waste fumes were recaptured and processed with the slag. As a result, almost total recovery was possible (Calvin and Bain 1900). Langworthy (1854–1855) states that five blast furnaces were operating, each processing 75,000 pounds of ore per week.

Dry bone zinc ore was roasted at 800°C, which drove off the carbonate and reduced the zinc to metal. Since the primary uses for zinc required the oxide form, the metal was subsequently oxidized to ZnO. The first smelting of blende ore was done in 1852 by Mattheisen and Hegler to produce spelter (zinc metal). High fuel costs required to recover sulfur created problems for early smelters. Most smelting of Iowa zinc was done in Wisconsin (Calvin and Bain 1900).

Production

Though records of early mining are sketchy and incomplete, is it estimated that as many as 500 mining operations were established in the Dubuque area after the lands were opened for development (fig. 1.8). A significant portion of these mines have been explored and mapped by members of the Iowa Grotto, a local chapter of the National Speleological Society. They estimate that miners sank between 700 and 2,000 shafts and drove 150 kilometers of tunnels in search of lead (Ludvigson and Dockal 1984). Production of lead from the Upper Mines (including mines in Iowa, Wisconsin, and Illinois) was greatest in the years 1845 to 1847, peaking at about 25,000 tons per year (Hall and Whitney 1858). After 1900 production of both lead and zinc declined sharply, except for a three-year period from 1905 to 1907 when production increased due to unusually high market prices. The last mine closed in 1910; in that year the combined annual production of lead and zinc concentrates totaled less than 270 tons. The industry's last gasp came between 1911 and 1917 when mine tailings and dumps were reworked. A brief but unsuccessful attempt to revive the industry was made in the 1950s (Ludvigson and Dockal 1984).

Although most of Iowa's lead production came from mines in the Dubuque area, lead was mined, albeit briefly, from two areas in Allamakee County. The Mineral Creek Mines, 10 kilometers north of Waukon, were developed along a branch of Mineral Creek during the 1850s. Galena and associated minerals, all highly oxidized, occur as scattered pockets in tough, chert-bearing dolostone; consequently, they were difficult to mine. Several openings were made into the hillsides along Mineral Creek. Individual drifts were 20 meters or less in length (Garvin 1982). Because of the scattered nature of the ore and the hardness of the rock, the mining lasted just a few years. Total production was only about 100,000 pounds of galena concentrate (Calvin 1895; Leonard 1896).

In 1892 lead was discovered near the village of Lansing in northeastern Allamakee County. Lead ore, consisting of galena and cerussite, was concentrated in a north-trending vertical fracture, which was traceable for about 350 meters. It was mined to a depth of about 20 meters. From 1893 to 1895 the Lansing Mine produced about 500,000 pounds of galena and cerussite concentrates (Calvin 1895; Leonard 1896).

Minor occurrences of lead-bearing mineralization have been reported at several locations in eastern Iowa (Heyl et al. 1959). None of these "diggings" have proven to be of commercial value.

A Legacy of Mining

Mining in the Dubuque area occurred long before the "age of environmental awareness." When mines played out they were simply abandoned.
Shafts remained open until surface materials caved in to plug the holes.
Eventually, timbered supports in shafts and tunnels rotted and collapsed.
The result was episodic subsidence of the land surface into underground
open space. It continues to be a problem for Dubuque residents. Subsidence generally occurs without warning, as exemplified by the collapse of
part of Hill Street on November 12, 1983 (Ludvigson and Dockal 1984).
Predicting where future subsidence might occur is made difficult by the
fact that the locations of many underground workings are not known with
certainty, since maps and other records were poorly kept. Plugging known
sinkholes is difficult, if not impossible, because the collapse may connect
to underground passageways that may extend long distances in unknown
directions. Plugs are often temporary and may collapse without warning.

Limestone

Iowa's limestone industry is by far the largest mineral industry in the state.
Limestone is quarried for use as building stone, crushed for aggregate,
and ground for use as a soil conditioner and for the preparation of portland
cement. Limestone and its close relative, dolomite (or dolostone), lie at or
very near the land surface in the eastern half of Iowa; consequently, most
commercial production has come from this area of the state.

Following the Black Hawk Purchase of 1832, the lands of what would
soon become the Territory of Iowa were being prepared for settlement by
Euroamericans. Early explorations by such men as Albert Miller Lea
(1835), John Plumbe Jr. (1839), David Dale Owen (1839), and John B.
Newhall (1846) revealed the presence of extensive outcroppings of limestone along the Mississippi River and along the major rivers of Iowa's eastern interior (Petersen 1957a). By the time of James Hall's 1855 survey of
the state, many quarries were already producing limestone, primarily for
building stone. Hall noted that a limestone quarry at Davenport supplied
much of the stone used to construct the first Mississippi River bridge at
Davenport (Hall and Whitney 1858).

Dimension Stone

In Iowa quarrying of limestone for building (dimension) stone dates back
at least to 1840. Fort Atkinson in Winneshiek County, built in that year, is
constructed of limestone. Foundations of early houses and barns were
limestone. It was also used for lintels, veneer, sidewalks, and flagging.
With the coming of the railroads in the 1850s, the demand for limestone
for bridge piers and ballast increased dramatically. Between 1859 and 1894
more than 150,000 railroad cars carried Iowa limestone to bridge-building

sites in Iowa and its six bordering states, and even as far away as the Pacific Northwest (Petersen 1957a).

Prior to the advent of mechanical equipment, removal of blocks of limestone began with the drilling of a line of holes along a cleared surface. Drilling was done by hand with hammer and chisel, and the depth of the hole was determined by the thickness of the limestone bed. Upon completion of the line of holes, the rock was split with a wedge and tilted away from the outcrop, where it was broken into smaller masses. After 1900 steam-powered drills and compressed air hammers replaced hand tools. Blocks were trimmed, sized, and then hauled by team and wagon until the arrival of the railroads.

The best known of the early stone quarries include the Bealer Quarry, which opened in 1883 along the Cedar River southwest of Tipton (fig. 4.24); the State Quarry in Johnson County, which supplied most of the limestone for the Old Capitol in Iowa City; and the Le Grand Quarry in Marshall County near the town of the same name. Sited in limestones of Mississippian age, the Le Grand Quarry is known for its magnificent crinoid fossils. Collected by B. H. Beane, many fossil slabs reside in museums in the United States and elsewhere. The State Historical Society of Iowa's mu-

4.24. Bealer Quarry, Cedar County, ca. 1900. *State Historical Society of Iowa, Iowa City.*

4.25. Stone City Quarry, Jones County, 1895. *State Historical Society of Iowa, Iowa City.*

seum in Des Moines houses an excellent display. The quarries at Stone City west of Anamosa have produced building stone for more than a hundred years from the Silurian Anamosa Dolostone (fig. 4.25). Distinctive because of the fine-scale laminations, building stones from Stone City can be observed in several prominent buildings in Iowa, including Botany Hall (Iowa State University), King Chapel (Cornell College) (fig. 4.26), and St. Martin's Church (Cascade). Dimension stone quarried from the Ordovician Galena Dolostone in northeast Iowa appears in the stone bridge at Elkader (fig. 4.27) and the Julien Dubuque monument near Dubuque. The old Humboldt School at Humbolt is made from Mississippian-age St. Louis limestone (Petersen 1957a). The development of portland cement and the high cost of producing dimension stone have severely reduced this part of the limestone industry in Iowa.

Chemical Lime

Long ago it was discovered that limestone could be "burned," that is, heated sufficiently to drive off carbon dioxide to produce calcium oxide, or quicklime. When water is added to quicklime, the calcium oxide recrystallizes to solid calcium carbonate. Quicklime was once used extensively for

4.26. King Chapel, Cornell College, Mount Vernon, Linn County.
Cornell College Office of College Communications.

building mortar, and the first lime plants in Iowa were built in the early 1880s. Gladfelter, Sugar Creek Lime, and Joyner Lime Company, all in Cedar County, were some of the earliest producers of quicklime. With the development of portland cement, which is more durable and has a longer setting time, the market for lime mortar gradually disappeared. Some quicklime is still produced by Linwood Mining and Minerals at Buffalo in Scott County for use in various applications by the chemical industry (McKay 1994) (fig. 4.28).

The chemical qualities of limestone are also important to its use in agriculture. Agricultural limestone, or ag-lime, is an effective soil conditioner because it reduces the acidity of highly acidic soils and provides calcium

4.27. Bridge crossing the Turkey River. Elkader, Clayton County.

4.28. Modern chemical lime plant. Buffalo, Scott County. *Linwood Mining and Minerals Corporation.*

as a nutrient. The ag-lime industry developed after World War II and continues strong today.

Crushed Aggregate

By far the greatest quantity of limestone produced in Iowa is crushed to make limestone aggregate. It is Iowa's leading mineral commodity, with an annual value of nearly $200 million (McKay 1994). Much of it is used in the surfacing of unpaved roads, for shouldering along paved roads, and for bases under hard-surfaced roads. Limestone aggregate that meets state-regulated wear and strength specifications is used in concrete. Aggregate makes up about 60 to 80 percent of the volume of concrete; the rest is lime or asphaltic cement. The strength and durability of the concrete depend upon the physical character of the aggregate and the degree of cementation of the aggregate particles. Concrete typically contains two types of aggregate—coarse and fine. The coarse aggregate is most often limestone, whereas the fine aggregate is generally river sand and gravel, which consists mainly of hard silicate materials such as chert, basalt, and quartz.

Quarrying is the preferred method of limestone extraction. In many areas of Iowa overburden consists of thin, poorly consolidated glacial and fluvial sediment. In the early days of quarrying, overburden was removed by horse-drawn scrapers, more recently by power shovels and scrapers. Where overburden is strongly indurated bedrock, the material is drilled and blasted. In cases where the overburden is too thick to remove by surface methods and where the quantity and value of the limestone warrant it, underground mining methods may be employed. Mines usually develop from preexisting quarries. Ranging in depth from 25 to 120 meters below surface, they may be of the drift, incline, or shaft type, and they typically operate by the room and pillar method. Over 300 limestone quarries currently produce limestone aggregate in Iowa, whereas only about a dozen underground mines are currently in operation (McKay and Bounk 1987).

To be suitable for marketing, limestone must be crushed and screened. Washing may also be necessary to remove quarry dust and clay. Before there were railroads, haulage was by wheelbarrow or by team and wagon. With the arrival of the railroads, tracks were laid in the floors of quarries, and rock was loaded by hand or by power shovel. Today, most limestone is loaded by power shovel or end loader onto large trucks or conveyor belts for transport to the crushing plant.

The crushing plant consists of a system of crushers, screens, and conveyors. In some quarry operations permanent crushing plants are built. For smaller operations the plant is portable so that it may be moved from one quarry to another. Screened rock is stockpiled by size and grade and then sold by the ton.

The need for hard-surfaced roads increased with the expansion of auto-

4.29. Lincoln Highway seedling mile sign.

mobile use after 1900. The first rural concrete-surfaced road in Iowa was laid near Eddyville in 1908 (Wood 1934). The first permanent concrete-surfaced road, the so-called seedling mile, was laid on the old Lincoln Highway between Cedar Rapids and Mount Vernon in 1918–1919 (fig. 4.29). During the Great Depression, reduction in demand for road-building material was partially offset by increased demand for limestone for Works Progress Administration (WPA) projects. The post–World War II economic boom brought a great increase in highway building, which continues today because of the installation and maintenance of the interstate highway system. At present, limestone is quarried in two-thirds of Iowa's counties (fig. 4.30), with an annual production of nearly 45 million tons (fig. 4.31).

Limestone is also used to line the banks of streams, lakes, and reservoirs as a means of controlling erosion. It is used to fill Gabian baskets, which are wire-mesh containers placed along road cuts and other unstable areas to reduce erosion and siltation.

Because large volumes of aggregate are required for all road construction projects, it is essential that local sources be available. Iowa is in a good position in this regard because large areas underlain by limestone are still accessible. However, urban growth and other competing land uses are already threatening access to aggregate sources in other areas of the United States, for example, in Chicago, where existing quarry expansion and new quarry development are severely limited by urbanization. Comprehensive land-use planning in Iowa is essential to insure that adequate local supplies of limestone, as well as sand and gravel, are not rendered inaccessible by urbanization.

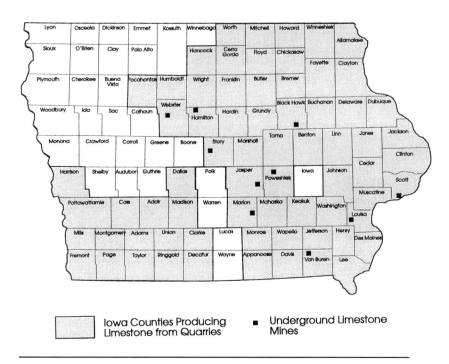

Lyon | Osceola | Dickinson | Emmet | Kossuth | Winnebago | Worth | Mitchell | Howard | Winneshiek
Allamakee
Sioux | O'Brien | Clay | Palo Alto | Hancock | Cerro Gordo | Floyd | Chickasaw
Fayette | Clayton
Plymouth | Cherokee | Buena Vista | Pocahontas | Humboldt | Wright | Franklin | Butler | Bremer
Webster
Black Hawk | Buchanan | Delaware | Dubuque
Woodbury | Ida | Sac | Calhoun | Hamilton | Hardin | Grundy
Monona | Crawford | Carroll | Greene | Boone | Story | Marshall | Tama | Benton | Linn | Jones | Jackson
Cedar | Clinton
Harrison | Shelby | Audubon | Guthrie | Dallas | Polk | Jasper | Poweshiek | Iowa | Johnson
Scott | Muscatine
Pottawattamie | Cass | Adair | Madison | Warren | Marion | Mahaska | Keokuk | Washington | Louisa
Mills | Montgomery | Adams | Union | Clarke | Lucas | Monroe | Wapello | Jefferson | Henry | Des Moines
Fremont | Page | Taylor | Ringgold | Decatur | Wayne | Appanoose | Davis | Van Buren | Lee

☐ Iowa Counties Producing Limestone from Quarries   ▪ Underground Limestone Mines

4.30. Limestone producers in Iowa. *After McKay and Bounk (1987), with data from the Iowa Limestone Producers.*

Cement

One of the important uses of limestone in Iowa is in the manufacture of cement. Cement is used to bind crushed rock, sand, or gravel to produce concrete and mortar. By the middle of the nineteenth century many Iowa towns had small lime kilns, which were used for the manufacture of mortar for building construction. Silurian dolostone was the preferred raw material because it is hard and durable and because its cement has a slow setting time. Although cements may be produced by a variety of processes, by far the most widely used generates portland cement. Portland cement was named by its British inventor, Joseph Aspdin, who patented the process in 1824. The stone he employed was limestone quarried from the Isle of Portland, a peninsula in southern England (McKay 1992).

Portland cement is produced by heating an artificial mixture of finely ground high-calcium limestone, silica, and clay to a temperature where partial fusion of the mixture occurs. The ratio is approximately one part silica/alumina to three parts limestone. Although a few localities in Iowa have produced "cement rock," that is, naturally occurring limestone containing clay in the right proportions, in most places limestone from one source is mixed with clay or shale from another. The limestone must be low in magnesium and free of chert. Good sources of cement limestone in

Iowa include the Platteville Limestone (Ordovician), which is abundant in northeast Iowa; the Otis, Wapsipinicon, and Little Cedar Formations (Devonian) in eastern Iowa; the Lime Creek Formation (Devonian) in northern Iowa; and several limestones of Mississippian age in central and southern Iowa (Eckel and Bain 1905) (fig. 1.3). The required clay is obtained chiefly from glacial deposits.

Limestone is quarried or mined by underground methods and hauled to the processing plant. There it is crushed, ground, and mixed with clay. Drying may be necessary to remove molecular water. Next, the mixture is preheated to about 870°C, at which point it enters the upper end of a slightly inclined rotary kiln, which is a large revolving drum 20 meters or

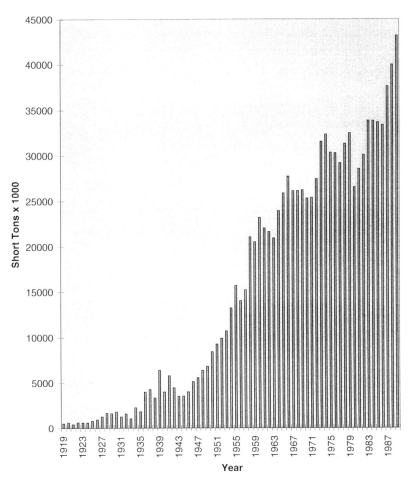

4.31. Crushed and dimension stone production in Iowa. *Data from the Geological Survey Bureau, Iowa Department of Natural Resources and the United States Bureau of Mines.*

more in length. The mixture is heated further to about 1370°C during its gradual transfer by gravity to the lower end of the kiln. Prior to the use of fossil fuels kilns were fired with wood. Several thousand cords of wood per kiln per year were required for continuous operation. Currently, coal and natural gas are the fuels of choice. The heating process drives off carbon dioxide and effects a series of chemical reactions involving calcium, aluminum, and silicon. The final products are calcium silicates and aluminates that are partially fused to glassy black nodules, generally referred to as clinker. The clinker is cooled, crushed, and ground to a fine powder. During this stage of the process 2 to 3 percent gypsum is added to retard the setting time of the cement. When the dry cement powder is mixed with water, heat-releasing chemical reactions occur, producing an interlocking mass of crystals. When combined with crushed stone, sand, or gravel, the result is tough, durable, water-insoluble concrete. The advantages of portland cement over other cements are that it will harden under water and that it has a high finish strength.

Portland cement was in use in Iowa by 1900. The first cement limestone quarry opened at Mason City in 1908. The demand increased dramatically in the late 1920s because Iowa, like many other states, was engaged in major road-building projects. Cement was used in building bridges and culverts, as well as for surfacing roads. Demand for cement in Iowa contin-

4.32. Modern cement plant. Buffalo, Scott County. *Lafarge Chemical Corporation.*

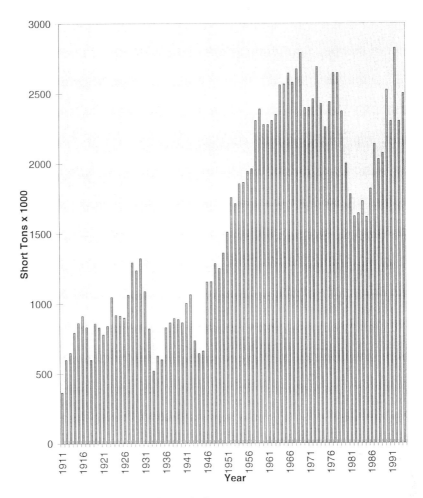

4.33. Portland cement production in Iowa. *Data from the Geological Survey Bureau, Iowa Department of Natural Resources and the United States Bureau of Mines.*

ues to be high today. Three plants produce cement from Iowa raw material: Lehigh Portland Cement Company and Holnam, Inc., at Mason City, Cerro Gordo County, and the Lafarge Corporation at Buffalo, Scott County. The Monarch Cement Company in Des Moines grinds cement clinker brought in from outside the state (McKay 1994). The Iowa cement industry currently accounts for a third of the total value of nonfuel mineral production in the state. In 1992 Iowa ranked tenth in production and total value among thirty-seven cement-producing states (McKay 1992). With the anticipated demand for cement in the construction and maintenance of Iowa's highways, and with large reserves of suitable limestone and clay in the state, the future of Iowa's cement industry appears bright (figs. 4.32, 4.33).

Peat

Peat has been used at least since Roman times as a source of fuel. The well-known peat bogs of Ireland are still a major fuel supplier; peat accounts for a third of the energy generated in that country. Peat results from the accumulation of partially decayed plant material. Areas of poor surface drainage, such as those found in regions of recent glaciation, are favorable for peat formation (Thompson 1992).

In Iowa interest in peat goes back to the early 1900s. Individual bogs up to 600 hectares in area and as much as 10 meters thick were identified (Beyer 1909). These bogs are confined to extreme north-central Iowa in the morainal area associated with the most recent (Wisconsinan) ice sheet. An example is the Goose Lake peat bog near Fertile in Worth County. Attempts were made to develop peat as a fuel source. It was soon discovered, however, that the high ash content (up to 25 percent) and the low heat value (about half that of Iowa coal) render Iowa peat not competitive with other fuel sources. Early tests of its suitability for use in the manufacture of paper revealed that its fiber content is insufficient to meet requirements (Beyer 1909).

The principal use of Iowa peat today is as a soil conditioner. In 1988 14,000 tons of peat were produced from two operations, with value of nearly half a million dollars. By comparison, in the same year 900,000 tons of peat were produced nationwide, with Florida and Michigan the production leaders (Thompson 1992).

Iowa peat might have future use as an alternative fuel source. In 1990 in Maine, the first peat-fired power plant, with a capacity of 22.8 megawatts per hour of power generation, went into production.

Future exploitation of peat in Iowa threatens a rich plant and animal ecology. Bogs are also good sites for the preservation of human and animal remains and cultural artifacts. Competing land use should be considered carefully in any decision about commercial peat development (Thompson 1992).

Petroleum and Natural Gas

In 1859 the first oil-producing well in the United States was drilled at Titusville, Pennsylvania. Not long after, the connection between commercial accumulations of petroleum and natural gas and sedimentary basins was established, which led to the discovery and development of the great oil fields of Texas, Oklahoma, Kansas, and Colorado. Oil and gas also began to be produced in states near Iowa, particularly in Kansas, Missouri, and Illinois. Increasing demand for petroleum products stimulated exploration in ever-widening areas.

The proximity of producing oil and gas wells to the borders of southwest and southeast Iowa and the knowledge that equivalent sedimentary rocks occur in Iowa have led people to believe that these hydrocarbons should also be found in commercial quantities in Iowa. To date, such accumulations have not been found. Since the early 1900s, the existence of small natural gas occurrences in southeast Iowa has been known. In fact, methane-bearing gas was tapped to provide light for nearby houses (Howell 1921). These small gas pockets were contained within the glacial drift overlying bedrock, but they were never large enough for commercial development. One hundred twenty-three known exploration wells have been drilled; three, all in Washington County, produced oil but in noncommercial amounts (Anderson 1992). Oil showings reported from time to time in most cases have been attributable to human activity. The presence of minute globules of bitumen in some southeast Iowa geodes and also in mineralized cavities along the Volga River in northeast Iowa is considered to be of scientific interest only. In 1963 a well drilled near Keota in eastern Keokuk County produced 400 barrels of crude oil over an eight-month period (Gilmore 1978; Anderson 1992). In oil parlance this occurrence would best be termed a "show."

In order for a reservoir of oil or gas to develop, several criteria must be fulfilled. A source rock for hydrocarbons must exist. Source rocks are commonly shales that are rich in partially decomposed organic matter. A reservoir must exist, that is, a rock containing sufficient permeability to allow hydrocarbons to accumulate must be present. Reservoir rocks are most often sandstones or highly fractured or cavernous rocks that otherwise would have very low permeability. Pathways that allow liquid and gaseous hydrocarbons to migrate from the source rock to the reservoir rock must exist. There must be an impermeable barrier that will prevent naturally rising oil and gas from escaping from the reservoir to the land surface, where they would disperse and decompose. The first two conditions are met in several areas in Iowa, particularly in the northeast and south (fig. 4.34). The third condition is probably met best in southwest Iowa in the area of the Thurmond-Redfield structure, which contains folds and faults that in adjacent areas of bordering states have provided good traps. The rocks of the Forest City Basin, the northern part of which extends into southwest Iowa, appear to meet all three requirements (Anderson 1983). The small producing field at Tarkio, Missouri, just 16 kilometers south of the Iowa border, is tantalizingly close (Anderson and Bunker 1982). Given the existence of suitable rock in Iowa and of producing wells near its borders, the lack of oil and gas in the subsurface is somewhat puzzling.

In 1987 Panamerican Petroleum Company, a subsidiary of Amoco Production Company, drilled a deep well just east of Halbur in Carroll County. Named the M. G. Eischeid #1, after the owner of the land that was leased for drilling, it bottomed at 5,441 meters, making it by far the deepest well

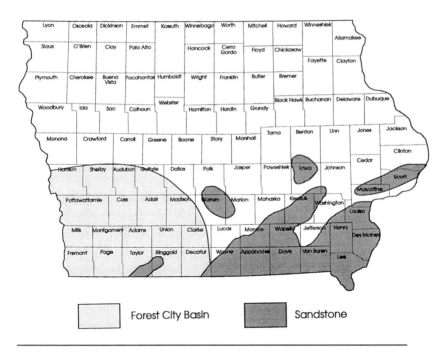

4.34. Distribution of potential petroleum resources in Iowa. *After Anderson and Bunker (1982).*

ever drilled in Iowa (fig. 4.35). The well was drilled as part of the exploration of a thick section of Precambrian clastic sediments that are associated with the Midcontinent Rift System (MRS). The MRS extends from southern Kansas to the northern shores of Lake Superior and passes through the heart of Iowa. The project was designed to test the premise that hydrocarbons might be trapped in the more porous horizons of the clastic sedimentary sequence. Traces of natural gas were found between 3,200 and 5,400 meters below surface, but oil was not discovered (Anderson 1990).

It appears that the prospects for Iowa becoming a significant producer of petroleum and natural gas are slim. The best chances are in the southwest part of the state, and any fields that might become productive will be small.

Sand and Gravel

Sand and gravel, together with crushed stone, constitute the aggregate industry. Sand and gravel are used extensively in surfacing gravel roads, as subgrade material in highway construction, and in concrete aggregate. In concrete, the aggregate is classed as coarse or fine. Coarse aggregate is most commonly crushed limestone, while fine aggregate is typically sand and fine gravel.

Sand and gravel deposits in Iowa are most abundant in active stream floodplains, terraces, buried stream valleys, and glaciofluvial areas, such as kames and eskers. Commercial quantities of sand and gravel are abundant along major rivers in central and eastern Iowa, such as the Des Moines, Iowa, Cedar, and Wapsipinicon. In the western and extreme northeastern parts of the state, good deposits occur in terraces. Buried sand and gravel also occur on glaciated uplands and may be exploitable where overburden materials are not too thick. Glaciofluvial materials on the Des Moines Lobe in north-central Iowa typically contain excessive fine sediment, but they may be locally useful (Kemmis and Quade 1988) (fig. 4.36).

Sand and gravel were first used in the manufacture of lime and plaster

4.35. Eischeid well. Photo by Ray Anderson. *Geological Survey Bureau, Iowa Department of Natural Resources.*

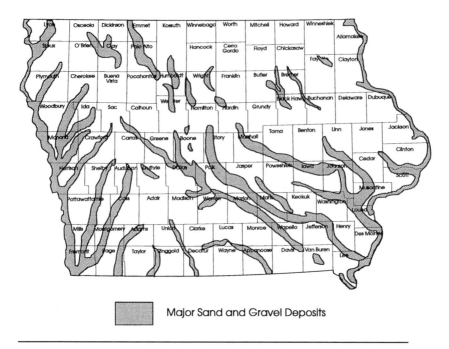

4.36. Distribution of major sand and gravel deposits in Iowa.

mortar. As with crushed rock, the development of road surfacing after 1900 greatly increased the demand for sand and gravel. Growth in production has been quite steady up to the present.

Because of their location at or very near the land surface, sand and gravel deposits in Iowa are extracted by surface methods. Before mechanized equipment, extraction was accomplished with a team-drawn scraper; currently, dredges and draglines are used. Pit material is then washed, screened, and stockpiled for sale as mortar sand, concrete aggregate, and ornamental stone (fig. 4.37).

Production of sand and gravel in Iowa is vigorous (fig. 4.38), and supplies appear to be adequate to meet the demand well into the twenty-first century. Because of high-volume demand and costs of transportation, local sources of supply are essential. Since good sources of sand and gravel occur along major rivers, they often underlie areas of urban and developing urban land use. Comprehensive land-use planning, with an eye to protecting supplies of sand and gravel from urban and other land use, is important in assuring a continuing supply of sand and gravel to meet Iowa's future needs for building construction and infrastructure maintenance.

Silica

Silica is a chemical industrial name for pure quartz sand. High purity is required for its use in iron foundries and in the manufacture of glass. Most natural sands and sandstones, though predominantly quartz, are impure and therefore unsuitable as sources of silica. However, if the sand has experienced several natural cycles of reworking, most of the impurities are removed by physical and chemical weathering. St. Peter Sandstone, named for St. Peter, Minnesota, where representative exposures of the rock can be observed, blankets a large area of the upper Midwest (fig. 4.39). The blanket ranges in thickness from a few to about 70 meters. Erosion by the Mississippi River has exposed a thick sequence of St. Peter Sandstone along picturesque bluffs in northeast Iowa and adjacent Wisconsin.

The high quality of St. Peter Sandstone was recognized by the early inhabitants of Clayton, a small town nestled against the Mississippi River bluffs in Clayton County, but mining did not begin until 1878. In that year William Buhlman opened a quarry along the bluff top at Clayton, from which sandstone was extracted and crushed and the sand sold to the Rock Island Glass Company for glass making. The sandstone, which is quite friable, was blasted and broken up with picks, loaded into gunny sacks, and hauled to the train depot. As demand for silica increased, a wooden trough and holding bin were built. The sand was washed down the trough to the holding bin, which was constructed so that sand would flow from it into railroad cars on a siding. Customers included glass factories in Milwaukee and iron foundries in several locations (Bischoff 1979).

4.37. Modern sand and gravel operation near Anamosa, Jones County.

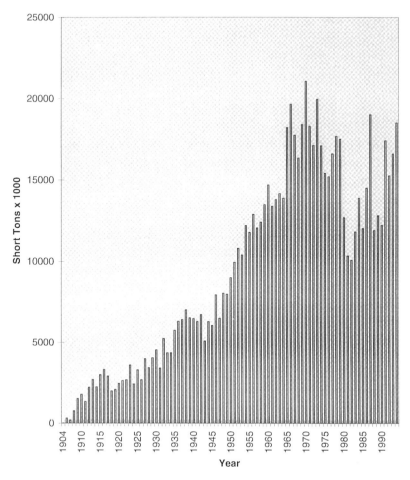

4.38. Sand and gravel production in Iowa. *Data from the Geological Survey Bureau, Iowa Department of Natural Resources and the United States Bureau of Mines.*

In 1905 Julius Buhlman (William had sold out to his brother some years before) sold the quarry for $3000 to Victor Drumb, who mined the sand hydraulically. Drumb leased the operations to the Clayton White Sand Company, whose main office was in Milwaukee. A crusher was installed in the pit to facilitate the breaking up of the sandstone. Production averaged about one railroad car per day. Mining continued under the managership of Charlie Blake until 1929, when competition and the Great Depression forced the operations to close (Bischoff 1979).

In 1920 Richard Kohl started the Clayton Brick and Tile Company. Using German machinery and technical expertise, the company produced brick, floor tile, and sewer tile. Sand for the brick plant was extracted from the bluff with picks and explosives. It was then loaded into a bucket suspended from a 28-meter-high mast, which was swung to a conveyor that carried the sand to the plant below. The brick was designed for backup and not for outside facing, but it was used to face several buildings in the local area, including the Clayton Town Hall and churches in West Union and Guttenberg. One of the first buildings constructed after the plant opened was a house in Clayton made entirely of bricks and tiles (most of them rejects) from the plant. Red tiles comprised the roof, and floor tiles were used in the basement.

In 1925 the brick plant was sold to Topet Taylor Engineering Company

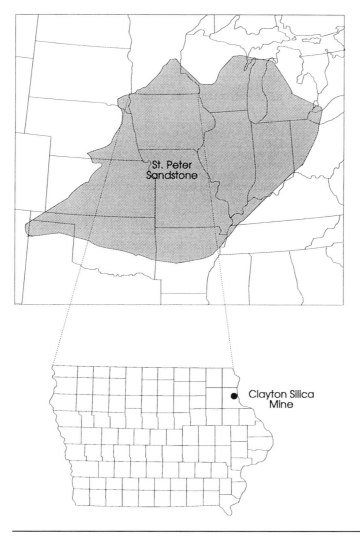

4.39. Location of the Clayton Silica Mine and the areal distribution of the St. Peter Sandstone. *Modified from Mintz (1979).*

of Pittsburgh, and the name was changed to the Iowa Brick and Slate Factory. Poor management and embezzlement of funds by the plant manager forced permanent closing of the plant in 1927 (Bischoff 1979).

The third mining operation at Clayton began in 1916, when John Langworthy, a chiropractor from Dubuque, formed the Langworthy Silica Company. It was his intent to develop a silica glass industry in northeast Iowa. Because of Langworthy's untimely death in 1919, the company's board of directors elected Otto Lange as chairman. The operation, which was run by Otto's son, Harvey, employed eight to twenty men, fluctuating with seasonal demand, and it shipped an average of 5 or 6 train carloads of wet sand per day. Production was severely reduced during the Depression (from 1,200 to as few as 40 carloads per year), but the company survived (Bischoff 1979).

Sand from the Langworthy Silica Company was marketed primarily for use as coring and molding in iron foundries. A major customer was the John Deere Tractor Works. Fine-grained sand was also used in finish plaster and for cutting and polishing marble. Because of heavy debt and wage disputes, the Langworthy Silica Company closed in 1941, and the mine remained inactive until 1945. At the urging of Deere, Concrete Materials, from whom Deere had been purchasing its coring sand, renewed the mining operations. Because of the high cost of removing limestone overburden, the company initiated room and pillar underground mining. The rooms were 12 meters square and the pillars 15 meters square. Ceilings were as high as 15 meters. The sand was drilled, blasted, and hauled by truck to crushers. As operations penetrated farther from the openings, a primary crusher was installed underground, and the product was carried by conveyor outside the mine for further crushing and screening. Because the operation was underground, mining could be performed year-round. Most of the sand was sold for foundry or industrial uses, with the John Deere Tractor Works in Dubuque and Waterloo as the main customers (Bischoff 1979).

During the Cold War in the 1960s, the Clayton Silica Mine was designated as a nuclear defense shelter. Ten railroad cars and ten semitrailers filled with food were placed in the mine, an amount calculated to support the human capacity of the mine, which was estimated at 44,000 persons. Since the population of Clayton was 500 and the population of the entire county was only 18,000, one wonders how 22,000 people would have gotten to the mine in the event of a nuclear disaster, let alone have found places to park their vehicles (Bischoff 1979).

Martin-Marietta Corporation purchased Concrete Materials in 1959 and continued operations at the Clayton Silica Mine until 1982. In that year the mine closed and was sold to Pattison Brothers, Inc., of Fayette for use as a grain storage facility.

5. Stories of Iowa's Minerals

Contained in Iowa's rich history are accounts that involve minerals and mineral resources. Though in a technical sense most of the stories are about rocks, it will be remembered from chapter 1 that rocks are composed of minerals. Therefore, any story about Iowa's rocks is a story about its minerals. Some stories, like that of the Cardiff Giant hoax, are fairly well known; others, like that of the McGregor sand painter, are more obscure. These stories are documented accounts that bring minerals and people together with greed, intrigue, tragedy, and humor in most interesting ways.

The Cardiff Giant Hoax

It would likely be the consensus today that the most famous mineral in Iowa is the quartz geode, but in 1869 it would have been gypsum. That was the year the giant was unearthed near Cardiff, an obscure village near Syracuse, New York. How Iowa became connected with New York and what that has to do with gypsum is a most fascinating story.

In the summer of 1868 George Hull, a tobacco farmer and cigar maker from Binghamton, New York, and H. B. Martin, a resident of Marshalltown, Iowa, checked in at the St. Charles Hotel in Fort Dodge. They were very secretive about the nature of their business, but they were interested in gypsum. The gypsum industry in Webster County was in its infancy (the first processing mill would not be built until four years later) but that there were extensive gypsum beds near Fort Dodge was well known. What Hull and Martin wanted was a single block of gypsum measuring at least 3.6 x 1.2 x 0.6 meters. That something shady was afoot is clearly indicated by the fact that Hull came all the way to Iowa for the stone. Gypsum was readily available in his own state of New York and had supplied a thriving plaster industry with raw material since 1835. When asked for what purpose they wanted such a large block of stone, they said that it was to be Iowa's contribution to the Lincoln Memorial and that it was going to be displayed in New York (how these two stories fit together is a mystery). After their offer to purchase a block from the Cummins gypsum quarry was refused, they leased some adjacent land and employed a quarrier named Michael Foley to remove a block of the stated dimensions, for which he was paid $15. The estimated weight of the block was about 10,000 pounds. Hull procured the services of several men, a large army wagon, and four yoke of oxen, and they succeeded in loading the gypsum block. The block was so heavy that the oxen could not pull it, so they added a team of horses. After getting stuck several times, and having broken several bridges, Hull and Martin deemed it prudent to trim the block in order to reduce its weight. It was dressed down to a weight of 7,000 pounds and eventually reached the railroad at Montana, Iowa (Northwest Book and Job Establishment [hereafter cited as Northwest Book] 1870; Gallaher 1921; Dunn 1960).

The gypsum monolith headed east by rail toward New York but made a temporary, but essential, stop at 940 North Clark Street in Chicago. There, Hull and Martin employed the services of a German stonecutter by the name of Edward Burghardt who, using Hull as a model, carved the block into a giant human form, a relatively easy task because of the softness of the stone. The finished product lay 3 meters long and about 1 meter wide at the shoulders. The right hand lay across the abdomen and the left beneath the back. The legs were slightly contracted, the left foot resting partially across the right. This posture was designed to create the illusion that the giant had been in pain at the time of death. Using an ingenious device consisting of steel needles imbedded in a wooden mallet, the three men created pits to look like pores in the skin. After several unsuccessful attempts with a clay model to get the hair to look right, they decided to leave the head bald. When the carving was complete, the giant weighed a mere 3,000 pounds. The finish work involved rubbing the entire surface of the colossus with a sponge filled with sand and water in order to remove the marks of the chisel and then treating it with ink and sulfuric acid to give the appearance of great age. Hull personally spent many hours wielding the sponge. The giant was then fitted with an iron-bound wooden coffin, which was marked "finished marble" and freighted to one George Olds at Union, New York (Gallaher 1921; Dunn 1960).

The trip to New York was circuitous, as was the wagon ride from the railway station to the farm of William "Stub" Newell (Hull's brother-in-law), which was located near Cardiff (fig. 5.1). Explaining such a large and ponderous box to curious onlookers required creative imagination, but Hull and Martin evidently were equal to the task, for no one in the area suspected what was in it. Upon arrival at the Newell farm, under cover of darkness and in great secrecy, the giant was carefully removed from his wooden coffin, or more accurately the coffin was removed from the giant. He was lowered from the bed of the wagon into his freshly dug grave,

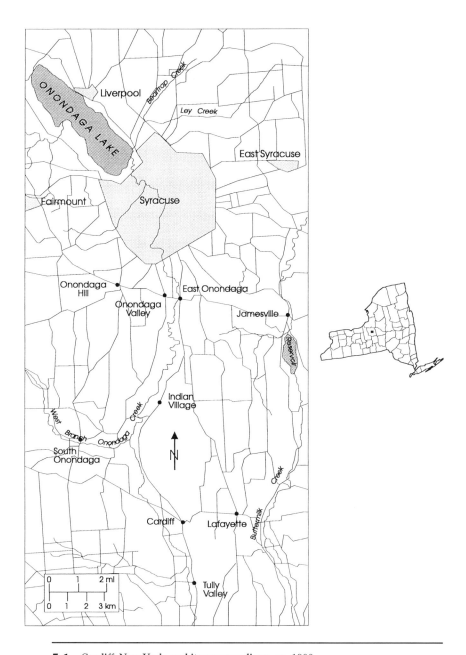

5.1. Cardiff, New York, and its surroundings, ca. 1900.

which was in a marshy area along Onondaga Creek at the foot of Bear Mountain. Burying the giant was not a small undertaking, considering the size and weight of the stone. As swamp water seeped in around him, the earth was packed tightly and the ground surface carefully smoothed to conceal any evidence of recent excavation (Northwest Book 1870; Gallaher 1921; Cardiff Giant, 1994).

A year passed. The burial site had been packed further by winter snows, which gave way to the growth of spring and summer vegetation. Nature had erased any remaining signs that the ground had been disturbed. On a Saturday morning on October 26 Gideon Emmons and Henry Nichols began digging a well at a spot designated by William Newell near the rear of his barn. At a depth of about a meter, a shovel struck a solid object. When attempts to pry it out revealed that the object was much too large to be easily moved, they proceeded, under the careful direction of Newell, to remove the dirt and mud from the object, and soon the petrified man looked up at the sky from a 3-meter-long muddy trench, whereupon Newell "innocently" remarked, "I declare, some old Indian has been buried here!" (Onondaga Indian tales of stone giants were well known in the area.) When it became clear that the stone had human form, word quickly spread throughout the neighborhood, and the curious flocked to the area to see the Onondaga Colossus, the name by which the giant was first known. Considering the stone man to be of great value, some speculators offered their farms, while others bid $10,000 or more in an effort to buy him. Newell shrewdly declined their offers, believing that he should wait and see how valuable his find might be. The first visitors were permitted to view the giant at no charge. On Monday, Newell went into business. He enlarged the trench and brought in a pump to keep out the water. He set up a tent around the gypsum giant to protect him from the elements, not to mention the eyes of nonpaying visitors, and he charged 25 cents a head. Guards were posted to protect Newell's investment. Two days later the *Syracuse Journal* announced the discovery. With increased publicity and demand to see the giant, the enterprising Newell raised the price of admission to 50 cents ("children half price—front seats reserved for the ladies"). Four-horse stagecoaches marked "To and From the Cardiff Giant" carried a steady stream of visitors at a dollar a head. Three thousand visitors came in one day. Business was so good that before long Newell could afford to hire a professional showman, Colonel Wood, to manage the exhibition. Newell turned a cowshed near the barn into a restaurant, and "The Giant Saloon" and "The Goliath House" satisfied the needs of thirsty patrons. Over the span of a few weeks, Newell is said to have taken in about $20,000 (Northwest Book 1870; Gallaher 1921; Dunn 1960; Cardiff Giant 1994).

Visitors to the site were awestruck by the size of the Cardiff Giant, the agony he must have suffered as indicated by the contorted legs and the hand on the abdomen, and the apparent great age of the stone, which was made all the more believable by the presence of deep grooves on the underside of the body, which they interpreted to be the results of water flowing beneath the body after burial. (In actuality, they were probably solution-enlarged fractures in the original stone.) The learned came from far and wide to make reasoned pronouncements on the age and origin of the giant. Medical doctors favored petrifaction as the logical explanation.

To explain the contorted position of the body, they proposed that he died of "cramp colic, brought on from eating too much sour krout [*sic*]" (Northwest Book 1870). Oliver Wendell Holmes and Ralph Waldo Emerson examined him and were greatly impressed with his antiquity and anatomical development. Some people thought he was buried recently, others anciently. Some thought the material to be Onondaga Limestone of local origin; others thought it was gypsum and foreign to the area. The contorted posture convinced some that an artist could not have sculpted it because, said they, it was not the sort of position an artist would choose. They were "impressed with the idea that they were in the presence of an object not made by mortal hand and that the figure before them once lived and had its being like those who stood around it. This feeling arises from the awful naturalness of the figure and its position. No piece of sculpture of which we have any account produced the awe inspired by the blackened form, lying among the common and everyday surroundings of the country farmyard" (unknown writer quoted in Northwest Book 1870). Perhaps the most distinguished investigator was James Hall, State Geologist of New York, who was, by interesting coincidence, formerly the State Geologist of Iowa. He correctly identified the stone as gypsum and thought the figure to be a statue, but he swore to the antiquity of the carving on the basis of the grooves and channels on the giant's backside which, he said, would require a very long time for the natural forces of water erosion to produce. In a written statement made on October 23, 1869, he declared that "the earth and sides of the pit bear no evidence of having been disturbed since its original deposition, and to all appearances, this statue lay upon the gravel when the deposition of the fine silt or soil began, and upon the surface of which the forests have grown for succeeding generations" (quoted in Northwest Book 1870).

While the debate about the origin of the giant continued, a steady stream of visitors flocked to the site to pay their hard-earned money to see what most of them believed were the remains of a petrified man. One woman claimed she could clearly see the blue veins in his legs. In fact, the bluish streaks, which are a common feature in Fort Dodge gypsum, looked liked human veins to Hull, and they convinced him that gypsum was the stone to use (White 1905). Myths and legends began to spring up, some no doubt created by Newell and his associates to increase visits to the exhibit.

Ten days after its discovery, Hull sold two-thirds interest in the giant to a five-man syndicate headed by a banker named David Hannum for the sum of $30,000, and the Cardiff Giant was moved to an exhibition hall in Syracuse. It took fifteen men and equipment to hoist it out of the ground. The price of admission increased to a dollar a person. P. T. Barnum sent an agent to Syracuse to view the exhibition, which on that day had attracted some 3,000 visitors. Upon receiving the agent's report, Barnum offered to purchase the monolith for $50,000. Hannum refused the offer. Sensing that money was to be made and not wishing to increase his offer, Barnum commissioned the carving of another statue, to which he charged admission under the same name—Cardiff Giant. Barnum billed his statue as the real giant and claimed that Hannum's was a fake. According to New York newspaper accounts, this state of affairs led Hannum, not Barnum, to utter the immortal words, "There's a sucker born every minute," in reference to the fact that people were willing to pay to see Barnum's fake (Brown 1993, personal communication). Despite competition from Barnum, the American Goliath, as it was advertised in Albany, continued to draw large crowds, as it also did when it was taken on tour around the country (fig. 5.2). Growing controversy surrounding its origin seemed to stimulate interest (Gallaher 1921).

Skepticism about the ancient origin of the gypsum giant began to increase. While it was still on the farm at Cardiff, Newell withdrew a large sum of money from the Onondaga County Bank for a draft to Hull (Dunn 1960). This raised some eyebrows about the relationship between Hull and Newell. Residents of Onondaga County began to remember seeing Hull with a wagon and a large box the year before. Hull and Newell attempted to quiet the growing concern by saying that the box contained heavy machinery and a quantity of contraband tobacco (hence the secrecy) (North-

THE GREAT
CARDIFF GIANT!
Discovered at Cardiff, Onondaga Co., N. Y., is now on Exhibition in the
Geological Hall, Albany,
For a few days only.

HIS DIMENSIONS.

Length of Body,	10 feet, 4 1-2	inches.
Length of Head from Chin to Top of Head,	21	"
Length of Nose,	6	"
Across the Nostrils,	3 1-2	"
Width of Mouth,	5	"
Circumference of Neck,	37	"
Shoulders, from point to point,	3 feet, 1 1-2	"
Length of Right Arm,	4 feet, 9 1-2	"
Across the Wrist,	5	"
Across the Palm of Hand,	7	"
Length of Second Finger,	8	"
Around the Thighs,	6 feet, 3 1-2	"
Diameter of the Thigh,	13	"
Through the Calf of Leg,	9 1-2	"
Length of Foot,	21	"
Across the Ball of Foot,	8	"
Weight,	2990 pounds.	

ALBANY, November 29th, 1869.

W. E. CASTLE, Printer, No. 59 State Street, Albany.

5.2. Billboard sign advertising the Cardiff Giant in Albany, New York. *From Dunn (1960).*

west Book 1870). Then, O. C. Marsh, a respected paleontologist from Yale University, examined the sculpture and determined that it was clearly of recent origin and an obvious fraud. The tide of public opinion began to turn against authenticity. Among the visitors to view the exhibit in New York was Galusha Parsons, a lawyer from Fort Dodge, who immediately wrote back to the local newspaper that he believed that the giant was carved from the great block of gypsum hauled out of Fort Dodge two years previously. It was not long before the connection between the Fort Dodge gypsum and the Cardiff Giant was firmly established. The owners protested vigorously, of course, that the statue was authentic, until Hull, who had already realized a substantial financial gain, publicly admitted the hoax (Gallaher 1921; Cardiff Giant 1994). Meanwhile Hannum brought suit against Barnum for calling his giant a fake. When word of Hull's confession reached the court, the judge ruled that Barnum could not be sued because, in fact, both giants were fakes.

With the exposing of the Great American Hoax, the Cardiff Giant's popularity waned. Copies of a thirty-six–page pamphlet published by some citizens of Fort Dodge and entitled "The Cardiff Giant Humbug" were sent to New York and distributed at exhibit sites (Northwest Book 1870). The pamphlet detailed the history of Hull's scheme, including his confession. Business really began to suffer. After traveling to Maine, Vermont, and Connecticut in the possession of C. O. Gott, the giant finally was stranded at Fitchburg, Massachusetts, in 1880 and was put in storage, where it remained for thirty-two years, except in 1901, when it was dusted off and displayed at the Pan American Exposition in Buffalo, New York. In 1903 Mark Twain featured the Cardiff Giant in a short story entitled "The Ghost Story." The giant was sold to Edwin Calkins for nonpayment of storage fees in June 1913. Joseph Mulroney purchased it later that year for the reported sum of $10,000 and returned it to Fort Dodge in January 1914. For the next few years Louis Mulroney took the giant on tour to several midwestern cities. It was then sold to Hugo Schultz of Huron, South Dakota, who displayed it in major cities on the West Coast. The Cardiff Giant Association purchased it and returned it to Fort Dodge in 1922, and it was exhibited for a time at the Hawkeye Fair Grounds. During the 1930s and 1940s the giant made occasional appearances at state fairs in New York and Iowa and for a brief period reportedly resided in the rumpus room of Gardner Cowles, a publisher from Des Moines. On May 19, 1948, the gypsum colossus was acquired by the New York State Historical Association and was moved to Cooperstown, New York, where it currently resides in the Farmer's Museum (Dunn 1960) (fig. 5.3). In 1972, in commemoration of the hundredth anniversary of the plaster industry in Iowa, a full-size replica of the giant was carved and put on exhibit at the Fort Dodge Historical Foundation's Old Fort. Currently, a movement is afoot to return the original sculpture to the place of its origin—Fort Dodge.

5.3. The Cardiff Giant at rest in the Farmers Museum, Cooperstown, New York. The carving is 3 meters long. *New York State Historical Association, Cooperstown.*

Even after the hoax became public, the Cardiff Giant still had believers. One was Alexander McWhorter, a graduate student in Hebrew studies at Yale, who announced that he had found unmistakable (to him) Phoenician inscriptions on the stone, which identified it as the god Baal. He published his theory about the "Phoenician Idol" in *Galaxy* magazine, a theretofore reputable publication (White 1905).

When asked why he had gone to such lengths to perpetrate a hoax on the American public, Hull responded that the idea came to him while visiting his sister in Ackley, Iowa. It seems that during a conversation with a Reverend Turk, a Methodist revivalist, there arose a disagreement over a passage in Genesis, "There were giants in the earth in those days." Hull, an avowed atheist, considered it another example of biblical fiction, believable only by the naive. He created the hoax to demonstrate that people gullible enough to believe Bible stories would believe in a modern-day giant. For a time he was right (White 1905; Gallaher 1921).

Soon after Hull confessed his hoax, he set about to fabricate another petrified man. This one he made of clay baked in a furnace, and it featured an internal skeleton and ape like tail and legs. The clay man was buried and subsequently unearthed in Colorado. Hopes for another financial success were dashed when his old nemesis, O. C. Marsh, appeared on the scene and quickly discredited the find (White 1905).

For those interested in further information concerning the Cardiff Giant, he has his own web page: http://www.cardiffgiant.com/hello.

Meteorite Impacts

That meteorites are extraterrestrial is a belief that has not been widely accepted until relatively recently. Until the beginning of the nineteenth century, the great minds of the world vigorously rejected the notion that rocks could fall from the sky. The numerous eyewitness accounts of fireballs and impacts were branded as folktales. Three hundred affidavits of a fall in southern France in 1768 were essentially ignored by the French Academy of Science (Wilson 1927). Thomas Jefferson was reportedly very skeptical of a meteorite fall in Weston, Connecticut, on December 14, 1807, and has been quoted widely as saying, "It is easier to believe that two Yankee professors would lie, than that stones would fall from heaven" (Wasson 1985). (In fairness to Jefferson, it should be pointed out that more recent research has failed to verify the quote.) A reason for this long-held skepticism is that scientists of those times did not make the connection between shooting stars and meteorites, because impact events are quite rare.

The passage of a meteor fireball through the Earth's atmosphere and its subsequent slamming into the ground is a landmark event that relatively few people in recorded history have had the fortune to witness. It is significant, therefore, that within a span of only fifteen years (1875–1890) no less than three meteorite impacts were seen by eyewitnesses within the borders of Iowa. Furthermore, these falls, in terms of total known weight of meteorite material, are among the eight largest in the United States and also among the largest in the world. The stories of the impacts and the discoveries of the meteorites involve awe-inspiring and terrorizing displays, greed, subterfuge, and even lawsuits. A brief summary of Iowa meteorites is given by Anderson (1993). Detailed accounts of each event are presented in the *Palimpsest* by Ben Hur Wilson, editor of *Earth Science* magazine and historian for the American Federation of Mineralogical Societies (Wilson 1927, 1928, 1929). An in-depth account concerning the Amana meteorites is given by Gustavus Detlef Hinrichs, a professor of chemistry at the University of Iowa and an avid meteorite collector and analyst (Hinrichs 1905). Summaries of some of these accounts follow.

The Marion Meteorites

The earliest recorded meteorite fall in Iowa occurred on February 25, 1847, southwest of Bertram, about 13 kilometers from Marion. Iowa was only sparsely settled, yet there were several witnesses. Most heard rumbling sounds followed by explosions—at least four separate blasts. Some heard whizzing sounds, as stone projectiles passed overhead. What was visible to observers during that afternoon hour was primarily smoke.

Judge James Cavanaugh and two of his sons were cutting wood along the Cedar River about 13 kilometers from the impact area. They described the sound of the passing meteor as humming and rushing, which increased to a roar. They also heard the whistling of what sounded like thousands of bullets in the air overhead. The roaring increased until the ground reverberated, culminating in a series of explosions to the northwest of their point of observation, which led them to believe that the village of Marion had been blown to pieces. That many of the local citizens did not appreciate what had happened is indicated by a report by some men that the rumbling and explosions were caused by a large rock that had fallen from a high bluff north of the Cedar River to the floodplain below.

According to C. W. Irish, a civil engineer, the meteor traveled in a northerly direction, passing directly over Iowa City. Several witnesses observed falling fragments. In one case, two workmen saw the snow fly about 70 rods from their position. Upon reaching the spot, they found that a meteorite weighing more than a kilogram had struck the frozen ground and bounced twice before coming to rest. Other fragments were found the following spring, the largest reported to weigh about 20 kilograms. Each stone was coated with a thin, dark brown fusion crust, inside of which the main body was a uniform gray. Some stones broke upon impact with the hard ground, exposing the gray interiors. Close inspection revealed the presence of iron sulfide and tiny blebs of Fe-Ni metal. The total number of sizable stones is not known with certainty. One reason is that some fragments were crushed by finders who mistakenly believed that the metallic grains were silver. The total known weight was about 28 kilograms (Wilson 1937). Two fragments of the Marion meteorite currently reside at the Field Museum of Natural History in Chicago, and one large fragment is at Tübingen University in Germany. The largest fragment (10 kilograms) is on loan to the University of Iowa and is on display in the Old Capitol in Iowa City (Ray Anderson, personal communication).

The Amana (Homestead) Meteorites

On the night of February 12, 1875, somewhere around 10:15 to 10:30 P.M., a fireball appeared in the southern Iowa sky. Eyewitness descriptions of the fiery projectile vary, partly because of the different locations of the observers in relation to its path and partly because some were in a state of shock and therefore could not give accurate accounts. The fireball was variably described as a round or pear-shaped object about 600 meters long and 125 meters wide, with the neck trailing behind. It was followed by a bright trail about 1,200 meters long and 15 meters wide. Its color ranged from brilliant white to dull red, while the trail was orange in the center, fading to green on the edges. Some witnesses observed flashes of red and green and sparks. Puffs of smoke or steam issued from the body of the meteor and trailed behind, somewhat like smoke from a moving steam locomotive. The brilliance was sufficient to eclipse the light from the moon and to create the momentary appearance of day. A combination of sputtering, hissing, roaring, and thundering sounds, described as like a high-speed train passing over a trestle, accompanied this dazzling display. Though the fireworks lasted but a few seconds, they were enough to send residents, livestock, and pets scurrying for the nearest cover (Wilson 1927).

The fireball was seen from Omaha to Chicago and from St. Paul to St. Louis. Its exact path is not known with certainty, but according to the most detailed description, made by C. W. Irish and reported by Gustavus Hinrichs in *Popular Science Monthly*, the meteor entered the atmosphere over the village of Mount Pleasant, in northern Missouri, and followed a progressively eastward-curving path successively over a point about 3 kilometers east of Centerville, to almost directly over Eddyville, to 2.5 kilometers east of Marengo. Its angle of descent was about 45° with respect to the land surface (Hinrichs 1905; Wilson 1927).

At the point where the meteor crossed the northwestern corner of Keokuk County it appeared to divide in two, the less brilliant portion traveling north-northeast and the far more brilliant portion continuing in a more northerly direction. The brighter portion exploded over Benton County with three terrific detonations, shaking buildings to the degree that it felt, to those in the nearest vicinity, as if an earthquake had occurred. Sounds of the explosion were heard as far as 120 kilometers away. To witnesses on the fringe of the path the sound on that bitterly cold February night was like the roar of a chimney fire, and many missed the dazzling light display in their haste to check their flues and fireboxes. The brighter, more northerly explosion apparently produced no recoverable meteorites. The fainter portion exploded in a meteor shower over Iowa and Amana townships in Iowa County, and it was this explosion that yielded the meteorites. These were observed by many as glowing coals falling from the sky, followed by the sounds of impacts upon the hard, frozen ground (Wilson 1927).

The first reported find was made three days after the fall by young Sarah Sherlock about 3 kilometers south of Homestead, while she was en route from school to the family farm. It weighed about 3.5 kilograms. News of the find sent people combing the immediate area, but additional fragments were not found. Finding meteorites, even when the general location of the meteor shower is known, is not an easy task, especially during the winter when the ground is covered with snow. On February 10 the area had been blanketed with a 40-centimeter snowfall. To further complicate the search, much of the area was wooded, and the meteorite was the stony type, which looks quite similar to some basaltic igneous rocks that had been transported into Iowa during the Pleistocene ice ages from the Lake Superior area. The great majority of the fragments were not discovered

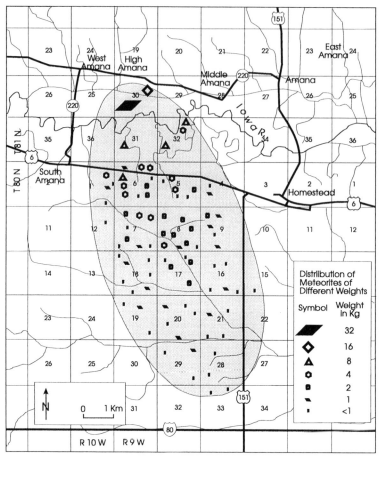

5.4. Location of the Amana meteorite impact. *After Hinrichs (1905).*

until the following April and May, most of these during plowing. A few were found in the woods by observing where they had struck and damaged tree limbs.

The meteorite field was approximately 11 kilometers long and 5 kilometers wide, with the long axis nearly north-south (fig. 5.4). Within the field the meteorite size increased generally in a northerly direction, reflecting the northward path of the meteor and the greater momentum of the larger fragments. The two largest fragments, one weighing 34 kilograms, the other 22 kilograms, landed in timbered river-bottom land about 1.5 kilometers west of Middle Amana near the millrace on the north side of the Iowa River, where they penetrated the frozen ground to a depth of more than half a meter (Hinrichs 1905). Hinrichs believed that, because the largest meteorites fell in bottomland, additional large fragments might have sunk in soft ground after spring thaw and not have been recovered, despite extensive searches. All told, about 230 kilograms of material were recovered. Based on close similarities in density and chemical composition among the fragments, the fragments were all part of a single stone (Hinrichs 1905).

Gustavus Hinrichs acquired a large collection (85) of Amana meteorites, many of which he sold to museums in the United States and Europe (data from Hinrichs 1905), including the Peabody Museum, Yale University; Museum of Natural History, Harvard University; U.S. National Museum, Smithsonian Institution; Paris Museum; British Museum, London; Naturhistorische Hof-Museum, Vienna; Royal Academy of Sciences, Berlin; Mineralogical Museum of the Imperial Academy of Sciences, St. Petersburg; Mineralogical Museum of the University of Copenhagen; University of Norway, Christiania; Riksmuseet, Stockholm; Academy of Sciences, Munich; Academy of Sciences, Brussels; Geological Museum, Lausanne, Switzerland; Museum, Harlem, Netherlands; and Hungarian National Museum, Budapest. Some of the meteorites are the property of the Amana Society, and some reside at the University of Iowa. A sizable fragment is on display at the Field Museum of Natural History in Chicago (fig. 5.5).

The fate of the largest fragment is worth recounting. Hinrichs (1905) quotes his own article, which appeared in the September 1891 issue of the *Chaperone*, of St. Louis. He fairly seethes with righteous indignation:

5.5. Several fragments of the Amana meteorites. *State Historical Society of Iowa, Iowa City.*

[The meteorite] was first "discovered" by a man, not a member of the Amana Society; he carried it home and thought he had found the best possible weight for his "kraut barrel." Just think of it, reader; a celestial visitor, degraded to its avoirdupois only, and serving to keep down the low fermentation in a barrel of sauer kraut! The Amana Society, learning that this magnificent specimen had been taken from their lands, forcibly liberated this celestial visitor from its state of degradation, and brought it back to their chief station, Amana. They kindly intrusted [*sic*] this specimen to my care and I removed it to my laboratory for study.

In the meanwhile the . . . traders made the "discoverer" bring suit in order to recover the weight for his kraut-barrel. The replevin proceedings failed to bring the celestial visitor to light; it was as if, in disgust, it had returned to its heavenly abode, however, this time, without attracting public attention by fire or thunder.

The suit progressed, experiencing all the beauties of American law. Some rich expert testimony was given, according to which this whilom kraut weight was worth a cool million of dollars. The pleadings were all that could be expected. The defendants argued that the heavenly visitor was as much a natural accretion to their lands, as rain or hail, or as much as "boulders" which had drifted to their land ages ago; that this stone in question, "coming with great dignity in a chariot of fire" was no less part of their land on that account, than if it had been dragged there on an ice floe before man had developed any taste for sauer kraut.

The court decided—in June, 1877, that this meteorite belonged to the owners of the land, the Amana Society. A couple of years later, the long missing specimen somehow made its reappearance at my house, where it remained in trust till called for by its rightful owners.

If the court had ruled against the Amana Society and Hinrichs, who refused to produce the meteorite, they would have been held liable for its value, which was set at its weight in gold (at least $20,000 in 1877). This was a test case, and it was used in the arbitration of lawsuits involving other Iowa meteorites.

The Estherville Meteorites

It seems that the excitement caused by the Amana meteorites had barely died down when another meteor announced its arrival in Iowa. The date was Saturday, May 10, 1879, and the place was Estherville, a small frontier town in the northwest part of the state. On this sunny spring afternoon the local residents were engrossed in a baseball game. Their first awareness of the meteor was the sound of a terrific explosion. By then the object had already passed, but in its wake trailed a long plume of white smoke, which stood out sharply against the almost cloudless sky. The track of the meteor was from southwest to northeast. From nearby vantage points, those witnesses who happened to be looking in the right direction at the right time saw a red streak. Several watched in utter amazement as the head of the streak exploded into a cloud of smoke that seemed to expand in every direction. Mrs. George Allen and her sister were out for a leisurely ride in an open buggy when they saw the explosion almost directly above their heads. From the ribbons of smoke that followed them they were able to identify three fragments that traveled separate curved paths in a northeasterly direction. C. W. Irish, the same engineer of Amana and Marion meteorite fame, happened to be locating a railroad line near the headwaters of the Des Moines River in southernmost Minnesota when he saw the fireball. A heavy thunderstorm was approaching from the west, and as he watched the ominous clouds and listened to loud reports of thunder the meteor burst through along the silvery edge of the cloud front. The head of the fireball was a brilliant white followed by a red streak. Irish, who was probably about 120 kilometers away from the impact area, listened for the sound of the explosion, but it was apparently drowned out by the thunder of the approaching storm. People in the vicinity of the explosion were greeted with jarring doors and windows and rattling furniture and dishes in cupboards. Some described the roar following the concussion like that of a tornado (Wilson 1928). The impact area was a few kilometers northwest of Estherville (fig. 5.6).

The first report of meteorite impacts came from a herdsboy, who ran breathlessly into the village of Superior exclaiming that he and his livestock had been peppered by small stones like hail. Several people observed the impact of the largest of the three large fragments. Dirt and debris were thrown high into the air in such a manner that the point of impact was quickly discovered on the farm of Sever H. Lee, a recent immigrant from Norway. Several neighborhood young men found the hole at the edge of a shallow slough. The crater was funnel shaped, about 3 meters in diameter at the top and somewhat elongated toward the northeast. Mud had been thrown up around the hole. Numerous fragments of metallic material were distributed raylike as far out as 90 meters from the crater. Several of the young men tried without success to dig out the mass, which lay about 4 meters below the ground surface. The sheer weight of the meteorite along with the swampy nature of the ground made removing it impossible. Not to be thwarted in their efforts, on Monday they secured the services of George Osborn, who with a block and tackle and windlass succeeded in hoisting the meteorite to daylight. It weighed 196 kilograms and measured 70 x 58 x 38 centimeters (Wilson 1928).

The second of the three large fragments was found four days later about 2.5 kilometers southeast of the largest fragment on the Amos Pingrey farm 3 kilometers north of Estherville. It was buried less than a meter deep and weighed about 68 kilograms (Wilson 1928).

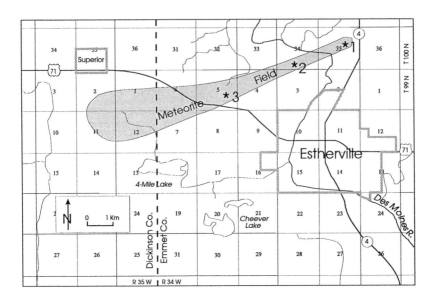

5.6. Locations of the Estherville meteorite impact. *Modified from Wilson (1928)*

The third fragment was not discovered until nine months later. The Pietz brothers, who were engaged in winter trapping in a swamp near the old Prairie Road, found a hole near the edge of the slough. Sounding the hole with a rat spear, they discovered a solid object about 1.5 meters down. This fragment weighed about 45 kilograms (Wilson 1928).

What happened to the three large meteorites? Meteorite 3 was quickly sold by the Pietz brothers for $100 to E. H. Ballard and George Allen of Estherville, who in turn sold it for the reported sum of $500 to Charles P. Birge, a lawyer and speculator from Keokuk (Henderson 1986).

In the case of meteorite 2, the owner, Amos Pingrey, apparently not appreciating its worth, gave it to a neighbor, John Horner, who concealed it in a cave on a neighbor's place. Gustavus Hinrichs, the central figure in the Amana meteorite story, having heard reports of the sensational meteorite impact, hastened to Estherville to see for himself. Meteorite 2 was produced from the cave, whereupon Hinrichs declared it to be an object of great value. Pingrey, dismayed at what he had given away, took steps to reclaim it. Horner retaliated by securing the services of Frank Davey, a lawyer and editor of the local newspaper, in order to defend his claim.

Meanwhile John S. Pilsbury, governor of Minnesota and a man greatly interested in meteorites, also having heard about the impact, sent E. J. Thompson, a professor from the University of Minnesota, to obtain as much of the meteorite as possible for the university's museum. Thompson, a shrewd businessman, appeared in Estherville with a sum of cash, a blank check from the governor, and George Chamberlain, editor of the newspaper in Jackson, Minnesota—who happened to be a good friend of Frank Davey's. At a midnight rendezvous, Thompson, Chamberlain, Davey, and Horner struck a deal for an undisclosed amount of cash. Thompson hauled the meteorite off to Minnesota without even having to use the blank check (Wilson 1928).

In the case of meteorite 1, the young men who recovered it hauled it into Estherville, where it was displayed in front of the public square for several days. Many people traveled to view it, some from a considerable distance. Recognizing its popularity, the young men got the bright idea that it could be put on display—for a price. They secured it in a strong box, loaded it into a wagon, and set out across Minnesota advertising the treasure on a sign that read:

> I am the Heavenly Meteor.
> I arrived on May 10th at 5 o'clock.
> My weight is 431 pounds.
> From whence I came nobody knows,
> but I am En Route for Chicago!!

During the excitement of removing the meteorite, consideration of who owned it was apparently forgotten. The young men had not gotten far when questions of ownership reached them. Fearful of losing their prize, they quickly returned and, wrapping the meteorite in an old quilt, quietly buried it on George Osborn's farm, carefully marking the location with four stakes. (Remember that the meteorite landed on the farm of Sever Lee, who apparently was never consulted before it was hauled off.) It remained buried well into the summer, the young men figuring that after the excitement died down they could unearth it again and resume their money-making venture. They unearthed it too soon (Wilson 1928).

The enterprising Charles Birge again appeared on the scene. He learned that the meteorite had been hidden away and also that Sever Lee was the legal owner. Lee had contracted to purchase a quarter section of land from the railroad, and it was on this section that the meteorite landed. Birge had somehow obtained the contract, which contained a forfeiture clause in case of default of payment. He discovered that Lee had neglected to make a payment on time; therefore, Birge was able to gain temporary possession of the land and with it title to the meteorite. Then he waited until the meteorite came out of hiding, whereupon he showed up with the local sheriff and instructed him to serve papers. He quickly spirited the meteorite out of the county before anyone could take counteraction. This act of piracy was never contested. Birge deeded the farm back to the Lees in October 1879 (Wilson 1928).

Birge quickly sold meteorites 1 and 3 to British and French museums at a handsome profit. Thus, with two of the meteorites in Europe and the third in Minnesota, Iowa was robbed of its treasure. Small metallic fragments continued to be picked up for several months. A favorite pastime for Estherville residents was meteorite picnics, during which hundreds of pounds of marble- to baseball-sized pieces were collected. Some of these were hammered into rings for residents by the local blacksmith. Other fragments went to museums within the state, to major museums in the United States—including the Field Museum of Natural History, the U.S. National Museum, and the Peabody Museum—and to museums in Budapest and Prague (Wilson 1928). A large polished slab and several small metallic fragments are on display at the Field Museum. A slice of meteorite 2 is on loan from the University of Minnesota to the Estherville Public Library (Henderson 1986).

The Forest City Meteorites

The last observed large meteorite fall in Iowa occurred late in the afternoon of May 2, 1890. With three events in fifteen years, meteor showers in Iowa were almost becoming subjects for weather reports. The sky was cloud free over north-central Iowa on the day of the event, and observers got a clear look at the fireball, which originated in the southwestern sky and passed rapidly overhead, leaving a trail of black smoke that hung in the sky for several minutes before dissipating. In fact, the smoke line could be seen from one horizon to the opposite. Vapors emanating from the smoke were described as sulfurous. The head of the fireball was brilliant white, and to some it appeared comparable in size to the moon. Its passage was accompanied by the usual roaring, sputtering, hissing, and thundering noises. The meteor exploded over Winnebago County, generating a large number of fragments that impacted very near the town of Thompson, about 18 kilometers northwest of Forest City (Wilson 1929) (fig. 5.7).

Witnesses were close enough to the impact site to begin recovery almost immediately. Some reported fragments landing virtually at their feet. Careful observation revealed that, despite the tremendous heat that was apparent in the gaseous envelope surrounding the fireball, the meteorites themselves were cold. Dry grass and hay in contact with even large fragments did not ignite; in fact, they were not even scorched. A geologist visiting the site determined that the clay in contact with one of the larger fragments showed no evidence of thermal alteration. The meteorite field was judged to be about 5 to 6 kilometers long and 2 to 3 kilometers wide.

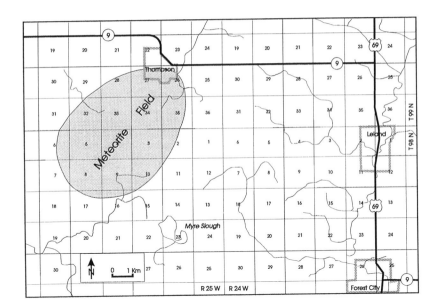

5.7. Location of the Forest City meteorite impact. *Modified from Wilson (1929).*

The two largest fragments weighed 40 and 30 kilograms. The rest (several hundred) were much smaller, most much less than a kilogram. The meteorites were the stony type and were identified by the presence of a blackened oxide crust. A Norwegian farmer named Hans Matterson reportedly brought in some fragments for display at a Forest City hardware store. He had broken them open, looking for silver. The silver specks seen within the interior suggest that the meteorite contained a small amount of nickeliferous metallic mineral. As usual, meteorite seekers swarmed over the field, and most of the treasure was recovered in a few days (Wilson 1929).

A local farmer, Peter Hoagland, reportedly found the 30-kilogram fragment and was delighted at the prospect of selling it for a considerable sum of money, which he and his wife intended to use to make a payment toward the construction of a new church. Seemingly out of nowhere, Horace V. Winchell, Minnesota's assistant state geologist, who was representing the University of Minnesota, appeared on the scene with a great interest in purchasing the meteorite. Before a deal could be struck, another prospective buyer arrived, and the bidding began. Winchell prevailed and presented the overjoyed Hoaglands with the tidy sum of $100. Winchell quickly loaded the meteorite into a strongbox and headed for town, stopping only long enough to purchase several of the smaller fragments from an eager farmer upon whose roof they had fallen. He deposited them at the Forest City railway station marked express for Minneapolis and awaited the early morning train (Wilson 1929).

Meanwhile, his unsuccessful competitor learned that the meteorite had landed not on the Hoagland farm but on the farm of neighbor James Elickson. Thus the stone belonged to Elickson (who in actuality was only a tenant), despite the fact that he had given Hoagland permission to remove it. A writ of replevin was quickly drawn up and served by the local sheriff, who confiscated the meteorite from the express office. It was placed in a vault for safekeeping until the courts could decide ownership. Peter Hoagland was forced to surrender the $100, church obligation notwithstanding (Wilson 1929).

The case was appealed to the Iowa Supreme Court, but before the case could be heard, the University of Minnesota posted a replevin bond (the court set the value at $500—five times the price paid the Hoaglands), took possession of the meteorite, and spirited it out of Iowa under cover of darkness. Once it was carried across the state line, it was thrown aboard a moving freight train, which carried it to Albert Lea, thence to Minneapolis by night express. It was then buried in a vacant shed, where it remained for two years (Wilson 1929).

In October 1892 the Iowa Supreme Court sustained the lower court's ruling, that meteorites belong to the owner of the ground on which they land. Thus they are treated like any mineral resource, despite their extraterrestrial origin. After the decision, the University of Minnesota was sued in an effort to retrieve the meteorite. The university claimed that it did not know the whereabouts of the stone. Consequently, the university was required by law to forfeit the $500 replevin bond. This it did readily and therefore was permitted to keep the prize, which mysteriously appeared soon after. At last report, the stone resides in a University of Minnesota museum. Once again, Iowans allowed a valuable meteorite to slip away to Minnesota. As with previous meteorites, fragments of the Forest City meteorite ended up in major museums in the United States and abroad and in private collections (Wilson 1929) (fig. 5.8).

The Mapleton Meteorite

Unlike the previous meteorites, the Mapleton meteorite was not a fall but a find, meaning that the time of its fall is unknown. Finds are most often rich in iron and nickel metal, and such is the case with the Mapleton meteorite. It was discovered by Harvey Meevers, who struck it at about 11:00 A.M. on June 17, 1939, while plowing a field a few kilometers east of Mapleton in Monona County (fig. 1.13). He concluded from its weight that it was a mass of iron, and for a while it served as a weight for his cultivator.

5.8. A fragment of the Forest City meteorite.

(Gustavus Hinrichs would have been mortified!) It might well have been discarded as a worthless lump of scrap iron had Meevers not happened upon an article in the *National Geographic Magazine*, that described the criteria for identifying meteorites. Convinced that his cultivator weight was a meteorite, Meevers displayed it as such in the window of a local bank. The meteorite measured about 40 x 25 x 15 centimeters and weighed 49 kilograms. How long the meteorite had lain in Meevers's field is not known, but the relative thinness of the coating of rust suggests a comparatively short time. That it might have been connected with fireballs sighted in the general vicinity in November 1916 or May 1917 is purely speculative (Wilson 1944).

After displaying the meteorite for a few weeks, Meevers contacted the

5.9. The Mapleton meteorite. *State Historical Society of Iowa, Iowa City.*

Field Museum of Natural History in Chicago with the hope that the museum might be willing to purchase it. Upon an initial expression of interest, he sent a small chip that had previously been sawed off. The museum received the chip on July 28, 1939. Museum officials soon verified that the mass was indeed meteoritic, and they arranged with Meevers to view it with the intent to purchase. Accordingly, Elmer S. Riggs, acting chief of geology, authorized Bryant Mather, associate curator of mineralogy, and Warren Raymond, assistant register, to meet with Meevers in Mapleton. Mather made an offer for an undisclosed amount. Meevers hesitated to consummate the deal on short notice, primarily because he had also offered the meteorite for sale to other interested parties and was looking for the best price. However, because other offers were not forthcoming, because the meteorite would be displayed in the Midwest, and because he was overcome by the $5 bills that Mather and Raymond kept laying out on the edge of his porch during the negotiations, Meevers agreed to sell. In addition to the monetary compensation, Meevers received copies of publicity about the Mapleton meteorite and membership in the Field Museum Society (Wilson 1944). The meteorite is on display in the Hall of Meteorites at the Field Museum of Natural History (fig. 5.9).

The Lost Creek Coal Mine Disaster

The worst mining disaster in Iowa history occurred in 1902 at the Lost Creek Coal Mine, about 16 kilometers southeast of Oskaloosa in Mahaska County. In those days, coal was mined by underground methods. Coal was broken out of a working face by drilling and blasting with explosives. On January 24, the blasting did not go as planned.

Just before noon Andrew Pash loaded two shots to break out coal for the afternoon's work. His room (No. 10) was located about 150 meters from shaft No 2. He had drilled one hole but decided to reuse a hole that remained from a previous firing. That explosion, instead of blowing inward to break up the coal, had blown out the tamping and had left the hole intact. Miners were heading to the surface for lunch, but about a hundred miners were still underground when Pash fired the shots. Instead of blowing into the coal, the shot was "windy," and this time the explosion set up a chain reaction. Fire shot from the hole and engulfed coal dust–laden air in the nearby rooms and tunnels, creating a horrific explosion. The fire and shock wave apparently also detonated barrels of blasting powder that were stored in the mine. The shock wave and a wall of fire swept toward shaft No. 2. En route they wrenched ventilation doors from their hinges, which rendered the ventilation system inoperative. Tracks were torn up and mine cars were blown into cages, the guides of which were torn away. All entry doors were blown out. Upon reaching the shaft the force of the blast threw timbers and debris upwards of 70 meters above the ground surface, disabling the mine cage and damaging the ventilation fan and other top works (Hoffman 1945; Booth 1990; Davis 1990).

Jasper "Jap" Timbrel, the mine superintendent, arrived on the scene minutes after the explosion, but four hours passed before the cage and fan were in working order and rescuers could descend into the mine. Workers threw buckets of water down the shaft in an effort to quench the fire. Men were driven back during early rescue attempts by suffocating gases. When they were finally able to access the east area, they came upon an awful scene of carnage. Several men were mangled and burned, some beyond recognition. Others were dead but showed no visible signs of injury; they were asphyxiated by damp (toxic gas released from the coal during mining). Rescuers, overcome by the toxic gases, were forced to ascend to the surface periodically to revive (Hoffman 1945; Davis 1990).

In all, twenty miners died at the scene, and fourteen were badly burned or injured. Ten mules also died in the explosion. Several of the dead men were foreign-born miners whose wives spoke little or no English, making communication of the tragic news all the more difficult. The disaster was blamed on faulty shot firing; in truth, the second hole should not have been reused. One week after the disaster, 1,500 miners left mines in the state to strike for better shot regulation. In April of that year a law was passed requiring that all shots be inspected before firing. It also stipulated that firing, which previously had occurred whenever a miner in a particular room was ready, take place only at the end of the working day, when most men were out of the mine (*Oskaloosa Daily Evening Gazette* 1902; *Des*

Moines Daily News 1902; *Cedar Rapids Evening Gazette* 1902; Hoffman 1945; Booth 1990; Davis 1990).

How the Geode Became Iowa's State Rock

Virtually every state has a state flower, a state tree and a state bird, and Iowa claims the wild rose, the oak, and the goldfinch, respectively. The idea of a state rock may seem strange, but in fact several states have adopted state rocks (California, gold; Ohio, flint; Michigan, coral (*hexagonaria*); Oregon, thunder egg; Illinois, fluorite; Rhode Island, bowenite; South Dakota, rose quartz; Florida, agatized coral) (Proctor 1967).

The idea for designating the geode as Iowa's state rock is believed to have been suggested by a visitor from South Dakota while speaking in Des Moines. The idea fermented until it found support in southeast Iowa. Nine state representatives from southeastern counties (where geodes are most abundant) spearheaded the drafting of House Resolution No. 14, which was introduced on the floor of the House on February 3, 1967. The resolution had bipartisan support and passed easily (93 to 26). The road to passage was much rougher in the Senate, where some senators considered it a triviality not worth the Senate's time. The debate was laced with sarcasm, with opponents wondering if the next resolution would promote a state animal or a state worm. Max Mills, a Republican from Marshalltown, suggested that the state adopt a state nut, and he implied that proponents of the geode resolution would be fitting candidates. The beauty of several geodes that were on display during the proceedings apparently turned enough hearts, for the Senate finally passed the resolution by a vote of 35 to 18, and the designation was made by the Iowa General Assembly (*New London Journal* 1967; Witkze 1987) (fig. 5.10). The wording of the resolution follows (Anderson 1993).

A JOINT RESOLUTION designating the Iowa geode as the official state rock for the state of Iowa.

WHEREAS, it is common practice for states to adopt specific flowers, birds, and trees as the official state flowers, state birds, and state trees, and

WHEREAS, it is also the practice among a number of states to adopt certain rocks as the official state rock of the state, and

WHEREAS, the state of Iowa does not at the present time have a rock as the official rock of the state, and

WHEREAS, Iowa has natural deposits of one of the rarest and most beautiful rocks in the example of the Iowa geode, and

WHEREAS, the Iowa geode is a much sought after brightly colored rock of a crystal formation and one of the finest geodes located in the nation, and

WHEREAS, Iowa is one of the few places where the geode formations are plentiful and found in some abundance, and

5.10. Iowa geode. Near Keokuk, Lee County. Specimen is 11 centimeters across.

WHEREAS, a survey conducted through the use of questionnaires mailed to rock collectors in the state has indicated that the Iowa geode is the first choice for the official state rock of Iowa; Now THEREFORE

Be it Resolved by the General Assembly of the State of Iowa:

1 SECTION 1. The Iowa geode is hereby designated and shall here-

2 after be officially known as the state rock of Iowa.

1 Sec. 2. The curator of the department of history and archives

2 is hereby directed to obtain samples of the Iowa geode adequate to

3 represent a fair sampling of the rock as found in this state and

4 display the samplings in an appropriate place in the state historical

5 library.

1 Sec. 3. The editor of the Iowa Official Register is hereby directed

2 to include an appropriate picture with an appropriate commentary

3 of the Iowa geode in the Iowa Official Register along with pictures

4 of the state flower, state bird, and state tree.

Iowa's Own Mineral—Iowaite

The State of Iowa not only has its geode, but it also has the distinction of having a mineral named after it. In the 1960s a mineral exploration program was conducted by the New Jersey Zinc Company in the Precambrian of Sioux County, Iowa. During examination of a diamond drill core obtained during the research, an unknown mineral was discovered in the depth interval from 300 to 450 meters. A thorough analysis of its physical and chemical properties and its crystal structure revealed that the mineral was a new species. It was named iowaite, in honor of the state where it was first reported (Kohls and Rodda 1967).

Iowaite is a very soft, platy mineral with a soapy feel. Its color ranges from translucent blue-purple to blue-green. On exposure to light the color quickly fades to dull gray or whitish green, and the texture becomes powdery (Kohls and Rodda 1967; Seifert and Brunotte 1996). The chemical formula is $[4Mg(OH)_2 \cdot FeOCl \cdot xH_2O]$, where $x \leq 4$. Subsequent work on the structure of iowaite by Allmann and Donnay (1969) indicates that $[Mg_4Fe(OH)_{10}]^+ \cdot [Cl(H_2O)_{x-1}]^-$, where $x = 4$, more accurately represents the chemical composition. The water is loosely held within the mineral structure, and its release, on exposure of iowaite to the atmosphere, accounts for the change in color and habit.

Iowaite occurs as veinlets in serpentinized ultramafic rocks in association with magnesium-bearing silicate and carbonate minerals. Seifert and Brunotte (1996) suggest that the mineral may form by the invasion of brucite $Mg(OH)_2$. by chlorine-rich fluids. Subsequent to its discovery in Iowa, iowaite has been reported from Siberia (Heling and Schwarz 1992), from the abyssal plain west of the west coast of Spain (Seifert and Brunotte 1996) and from Transvaal, Republic of South Africa (advertised by the Mineralogical Research Company of San Jose, California). The instability of iowaite under atmospheric conditions indicates that it is very unlikely that the mineral will ever be found at the surface.

The Ottumwa Coal Palace

The late nineteenth century in the United States might appropriately be referred to as the Age of Expositions. Expositions were the forerunners of world fairs and were held on national, state, and regional levels for the purpose of furthering the aims of agriculture, industry, and commerce. In Iowa expositions often centered around an edifice called a palace, which was a grand exhibition hall in which the fruits of agriculture and industry might be displayed. The first palace was built in Sioux City in 1887. A large and imposing structure, it was called the Corn Palace because it was decorated inside and out with corn. Sioux City's Corn Palace and festival were so successful that repeat performances were held annually from 1888 to 1891, each festival with a new or rebuilt structure. Influenced by Sioux

City's success, other Iowa communities soon created their own festivals, each with its own palace. For example, Creston had a Blue Grass Palace and Forest City a Flax Palace (Schwieder and Kraemer 1973).

Some time prior to the summer of 1890 Henry Phillips, Calvin Manning, and Peter Ballingall, all prominent Ottumwa citizens, determined that Ottumwa should have its own exposition and palace. The three men formed a company and sold stock at $5 per share. Investors were slow to respond, and for a time it appeared that the grand scheme would not get off the ground. Ballingall called a meeting of local citizens at which he gave a spirited speech and increased his own investment to $700. Before the meeting ended more than $30,000 had been pledged, an amount deemed sufficient to proceed with plans for the exposition (Kreiner 1992; Petersen 1963).

The plans moved forward in earnest, and in the summer of 1890 construction of the palace began on a lot near the Burlington Railway station. Since Ottumwa was in the heart of Iowa's coal district, the decision was made to feature coal mining and other industries as well as agriculture. Nine counties in the coal district (Appanoose, Jefferson, Keokuk, Marion, Mahaska, Monroe, Wapello, Warren, and Van Buren) were invited to display their wares (Petersen 1963).

To say that the Coal Palace was an awe-inspiring edifice would be an understatement. It was 70 meters long, 40 meters wide, and two stories high, and it boasted a central tower that was 60 meters tall and 11 meters wide. Combining Gothic and Byzantine architectural styles, it contained towers, battlements, domes, and turrets, which gave the palace a medieval character. In all, 150 pillars symbolizing coal shafts were used in the construction of the palace, and no two were alike. The first-floor exterior was veneered with blocks of coal set in black mortar; pea and nut coal coated the second-story walls. Vitric coal (which is highly reflective) set in red mortar adorned the turrets. This same coal was used to create the title, "Coal Palace." On a tower above the sign were two murals, one depicting the "Carboniferous Age," the other a coal mine with a dump, an engine house, a loaded coal train, and a miner wielding a pick. Balconies above the wings of the main building provided visitors with panoramic views of Ottumwa. Two second-story gables depicted the goddess Ceres surrounded by agricultural equipment and the god Vulcan with a factory in the background. The windows, designed to imitate cathedral glass, were illuminated at night with colored electric lights. Flags bearing the names of each of the participating counties flew atop the turrets and towers (*Ottumwa Daily Courier* 1890; Beitz 1962; Petersen 1963) (fig. 5.11).

Visitors entered the Coal Palace, which was also referred to as "King Coal's Castle" and the "Palace of Jet," through grand arches in the tower. The main floor contained an auditorium with a seating capacity of several thousand. Within the tower, 45 meters above ground, was an observation

5.11. The Ottumwa Coal Palace. *State Historical Society of Iowa, Iowa City.*

gallery and a dance pavilion. Inside the palace the pillars, walls, and rafters were completely hidden by exhibits of agricultural, mineral, and mechanical products. Corn, oats, wheat, rye, barley, millet, blue grass, timothy, clover, flax, and pumpkin and cucumber seeds were skillfully blended with red and blue fabric and paint in patterns of brilliant color. The interior walls contained pictures in corn that symbolized agriculture, industry, mechanics, music, art, literature, geography, and commerce. Rooms with impressive names like Flora's Temple, Ceres Palace, and Egyptian Temple, decorated with seeds of all kinds, housed county and other displays. Exhibits featured food industries, mining, manufacturing, jewelry, photography, and millinery, among others (*Ottumwa Daily Courier* 1890).

Directly opposite the entrance was a mountain scene featuring a stream of water that plunged almost 12 meters down from a backdrop of canvas mountains into a sculpted valley. Crags and boulders jutted out of the water, live fir and spruce trees grew in the valley, and a suspension bridge spanned it. The stream flowed into a waterfall that tumbled to a miniature lake 4 meters below. Calcium lights played on the cataract. It was a sight to behold. About 1.5 million gallons of water per day were required to keep the system running. In front of the cascade was a large platform for speakers, visiting dignitaries, and performing artists (Beitz 1962).

A unique feature of Ottumwa's Coal Palace was a miniature coal mine. Visitors entered a dark, coal-lined shaft from the gallery 45 meters above the main floor and then were lowered in a mine car to a level below the main floor where a mule-powered train of pit cars awaited them. They were hauled to simulations of coal mine entries, rooms, tracks, and miners demonstrating their craft. Rich seams of coal were visible along the "tunnel" (*Ottumwa Daily Courier* 1890). It is remarkable that the entire palace with all its furnishings was completed in just three months.

Prominent citizens, including government dignitaries, were sent invitations. On September 16 the exposition opened with a formal procession including Iowa Governor Horace Boies. Peter Ballingall opened the ceremony by announcing, "Karbon is King, and his palace is open to his subjects." The governor delivered an inspiring speech, and a chorus provided atmosphere by singing "Down in a Coal Mine." Six thousand people crowded into the palace on opening day. The exposition ran from September 16 to October 11. Nearly every day some organization, group, or geographic area was honored. There were days for Iowa, Missouri, Cedar Rapids, Des Moines, and each of the participating coal counties. Old Soldiers Day, Ladies Day, Miners Day, Labor Day, Railroad Day, and even Democratic Day and Republican Day were advertised, anything to draw more paying customers to the exposition. Bands and choruses entertained on the platform, and a comic opera was performed (*Ottumwa Daily Courier* 1890; Beitz 1962).

The climax of the exposition occurred on October 9 when President Benjamin Harrison paid a visit to the Coal Palace. The presidential party was taken on a tour of the palace and treated to a grand parade that reportedly was witnessed by at least 40,000 visitors. At the appointed hour the president was escorted to the platform to deliver a speech, in which he extolled the virtues of the common things of life. Regarding the Coal Palace, he said, "If I should attempt to interpret the lesson of this structure, I should say that it was an illustration of how much that is artistic and graceful is to be found in the common things of life and if I should make an application of the lesson it would be to suggest that we might profitably carry into all our homes and into all neighborly intercourse the same transforming spirit" (quoted in Kreiner 1922). At some point during the speech someone (Harrison suspected a Democrat) turned on the water cascade, the noise of which drowned out the speech. In the evening nearly 10,000 people crowded into the palace to shake hands with the president— someone estimated a rate of about sixty shakes per minute. It was an auspicious occasion, indeed (*Ottumwa Daily Courier* 1890; Petersen 1963).

Pleased with the success of the first exposition, the promoters decided to hold it again the following year. The Coal Palace was refurbished for the second exposition. The second Ottumwa Coal Palace was also the last; this venture was financially far less successful than the first, despite the fact that a new president, Grover Cleveland, as well as Congressman William McKinley, put in an appearance. Not insignificant to its demise is the fact that Peter Ballingall, chief architect and principal supporter, died suddenly in 1891 on a return voyage from China (Schwieder and Kraemer 1973). Not long afterward the Coal Palace was torn down and passed into history. The land on which the palace stood is now known as Ballingall Park and is a part of the estate Ballingall willed to the city (Beitz 1962).

The McGregor Sand Painter

When one thinks of Iowa artists, the name Andrew Clemens probably will not be on the list. Yet in the late nineteenth century his works of art were well known, not only in northeast Iowa but also in the bordering states and as far away as Europe.

Andrew Clemens was born on January 29, 1857, in Dubuque, Iowa, the son of German immigrants. His father and uncle owned and operated a wagon shop. Hearing that McGregor was a focal point for grain shipping and for immigrants who were making their way west, the Clemenses set up business on Main Street in McGregor in 1858. Andrew and his two older brothers were typical boys who enjoyed the excitement and activity of the river town (Rischmueller 1945).

In 1862 young Andrew Clemens contracted encephalitis, and the disease left him deaf. After several years of tutoring at home by his mother, he was enrolled in the Institution for the Education of the Deaf and Dumb at Council Bluffs, where he received a basic education and acquired woodworking skills. His summers were spent in McGregor, where he and his brothers explored the bluffs and ravines near his home. One of his favorite spots was Pictured Rocks, and it was here that he launched his career as an artist (Rischmueller 1945).

Pictured Rocks (not to be confused with a scenic area of the same name that is located on the Maquoketa River in Jones County) was well known to local residents. Located along the bluffs of the Mississippi River about 3 kilometers south of McGregor (in present-day Pikes Peak State Park) and opposite the Wisconsin River is a wooded ravine carved into the bluff. Beds of St. Peter Sandstone are exposed along the sides and bottom of the ravine. St. Peter is typically a clean, friable sandstone that is most often white to light gray in color. Locally, however, the quartz sand grains are coated with iron oxide, which imparts red, purple, yellow, and brown colors to the sand. The source of the iron oxide is believed to be the overlying carbonate rocks and shales (Leonard 1906). At Pictured Rocks the colors are vivid and highly variable, and the site became a favorite spot for picnickers. Andrew Clemens was fascinated by the multicolored sand layers, and during one summer vacation he collected sand, each color in a separate container. He then proceeded to pack sand into a colorless glass bottle, incorporating the different colors into a simple geometric design. In succeeding summers, as his skills improved, the designs became more intricate and the color blending more subtle. Local residents soon became aware of his work and offered to buy his "sand paintings" (Rischmueller 1945).

Clemens continued his studies at the school in Council Bluffs until February 1877, when a disastrous fire destroyed the main building. The school closed abruptly, leaving him at a crossroads. Impressed with his abilities, his instructors encouraged him to receive further training at the Smithsonian Institution in Washington, D.C., in order that he might return to the school as an instructor when it reopened. Instead, he returned to McGregor to continue his sand painting (Rischmueller 1945).

With the aid of his parents and brothers, Clemens obtained enough sand to last him for several months and a supply of bottles, and he began to paint in earnest. The tools of his art were simple: a long-handled tin scoop used to insert the sand into the mouth of the bottle, narrow sticks for shaping the sand into the designs he desired, and packing tools for pressing the sand tightly in place. The tools were homemade and the techniques self-taught. The sticks, or "brushes," were carved from green hickory and tempered over a candle flame. His "palette" contained shades of red, purple, orange, yellow, green, and even blue, as well as white, gray, black, and brown, some forty different colors in all (Downing 1973). All of the colors were natural, and all were obtained from the Pictured Rocks area, except for the white sand, which he got at White Springs west of town. Lacking sieves, he spread the sand on a piece of coarse-textured paper and gently rubbed the sand with the bowl of a spoon. He used only the small grains that adhered to the paper, and therefore the size was quite uniform. The paintings were backed with white sand, which was pressed tightly into the bottle to prevent the colored sand from shifting. Once the bottle was full, it was stoppered and sealed with pitch (Rischmueller 1945; Wythe 1953).

Aware of the quality of Clemens's work, Henry Goldschmidt, whose grocery was across the street, offered to sell his sand bottles on commission. Goldschmidt also provided a workspace in a lean-to that adjoined the grocery. Clemens artwork increased in popularity, and before long the demand exceeded the supply. Each bottle took from two days to two or three weeks to make, depending on the size and the complexity of the design. His paintings were purchased as gifts for steamboat captains, and they were bought by travelers passing through McGregor. As word of this unique art spread, he began to receive orders by mail.

After two years in Goldschmidt's employ, Clemens decided to go into business for himself. He set up a worktable at the front window of his parents' home, where his work could be viewed by passersby. His designs were self-created, except when he was commissioned to copy a picture. No written description can do justice to the uniqueness and magnificence of his artistry. His bottles contained miniature representations of a pontoon

5.12. The George Washington bottle, a sand painting by Andrew Clemens. The bottle is 11.5 centimeters in diameter and 30.5 centimeters high. *State Historical Society of Iowa, Iowa City.*

railway bridge built across the Mississippi River at Prairie du Chien, one of the first steam engines used on the Chicago, Milwaukee, St. Paul and Pacific Railway; a railway locomotive, less than 5 centimeters long, with the number 1406 inscribed on the cab using single white grains of sand each placed atop another; a sailing vessel on a white-capped sea, silhouet ted against white clouds in a mauve and orange sky, with a background of delicate geometric designs and bands of leaves and flowers (Risch-mueller 1945); and the Lord's Prayer and the Ten Commandments spelled out letter-by-letter (Wythe 1953). The George Washington bottle is considered his finest work. The round-top bottle was 30.5 centimeters high and 11.5 centimeters in diameter. It displayed "sand miniatures of George Washington astride his favorite white horse, a side-wheel steamboat correct in every detail, two American Indians in ceremonial regalia standing against a tepee, and an accurate reproduction of the Great Seal of Iowa. The State Motto, 'Our Liberties we prize, and Our Rights we will maintain' [was] a marvel of precision" (Rischmueller 1945) (fig. 5.12). What makes his work all the more remarkable is that most of the bottles he used were designed for display mouth-end-down; thus his paintings had to be created upside down.

As Clemens's fame spread, orders began to come from England, Germany, and other parts of Europe. Yet his prices were incredibly low—ranging from $1 for simple designs to $8 for a complex painting bearing the name of the owner. One bottle is reported to have sold for $35 (Downing 1973).

The work was tedius, demanding, and confining. Since he worked with dry sand, he likely inhaled a considerable amount of silica dust. The work took its toll, and Clemens contracted tuberculosis. He was invited by officials of the Columbia Exposition of 1893 to work and display his sand paintings in Chicago. However, his health was failing, and he was forced to decline the offer. In May of 1894 he passed away in McGregor at the age of thirty-seven (Rischmueller 1945).

The total number of Clemens's sand paintings is not known, but there were hundreds. His work can be seen at the State Historical Society of Iowa's museum in Des Moines (including the famous George Washington bottle), at the State Historical Society of Iowa in Iowa City, and at the Historical Museum in McGregor. Regrettably, most of his art has not survived because of the fragility of the glass containers and because he did not use adhesives to bind the sand together. If a bottle broke, the sand simply ran out.

The art of Andrew Clemens is unique. Using one of the most common minerals in Iowa as his medium, he perfected a most uncommon art form.

Appendix A

How to Identify Iowa's Minerals

Accurate mineral identification is an essential prerequisite to interpreting mineral origins. Minerals that have very different histories may look alike. In Iowa mineral deposits, pyrite has been misidentified as chalcopyrite, capillary marcasite as millerite, brown rhombohedral calcite as fluorite, and white calcite as strontianite. With the right identification tools, these minerals usually can be distinguished quite easily from one another.

Minerals are identified by their physical and chemical properties. The physical properties of a mineral are consequences of the chemical elements it contains and the strength and geometric relations of the chemical bonds between those elements. Chemical properties are also related to chemical composition and bond strength. Additional information about physical and chemical properties of minerals can be obtained from any physical geology or mineralogy textbook (e.g., Klein and Hurlbut 1993).

Color and Streak

Color in minerals, as in all substances, results from the selective absorption of visible light. The causes of color in specific minerals can be understood only if one is well grounded in principles of structural chemistry. Most often, color is related to the presence of certain chemical elements (called chromophores), whose electronic behavior causes selective absorption of light energy. Iron, manganese, and copper are common chromophores. Where chromophores are major constituents of a mineral, they impart a characteristic and predictable color. For example, malachite (chromophore = copper) is always green, and limonite (chromophore = iron) is always yellow brown. However, if the chromophore is a minor constituent or if the color is produced by some other mechanism, mineral color may vary from one occurrence to another. Calcite comes in virtually every color of the rainbow. In Iowa, yellow, brown, gray, black, pink, purple, and orange calcites are known, as are the more common colorless and white varieties.

Some minerals exhibit a zonal variation in color, which means that the core of the mineral may be a different color than the rim. Some fluorites from the Postville Quarry have yellow cores and purple rims. Zonal variation may be due to the presence of microscopic inclusions of iron sulfide minerals, as in the case of the zoned calcites from Robins Quarry (fig. 1.1).

Streak is the color of the mineral when it is powdered; it may be observed by rubbing the mineral across a piece of unglazed white porcelain tile or any slightly roughened surface. Most minerals that exhibit variable color produce white or colorless streaks. The majority of Iowa's minerals show white streaks. Exceptions are the metal oxides, like hematite and limonite, and the sulfides, like pyrite, galena, and sphalerite. Streak is more reliable than color in the hand sample. If the streak is colored, it is important to describe the color as precisely as possible. For minerals with dark streaks, the difference between greenish black and grayish black may be significant. Also remember that streak tests are destructive—when you rub a mineral across a streak plate you will invariably damage the mineral.

Some minerals display color when illuminated with ultraviolet light (UV). UV sources produce long-wave (365 nm) or short-wave (254 nm) energy. UV color results from chromophores in the mineral or from defects in its crystal structure. Minerals in Iowa that commonly exhibit UV fluorescence include some varieties of calcite, fluorite, and barite. When using UV, remember that the UV radiation is harmful; shield your eyes and skin from the radiation.

Luster

Luster refers to the manner in which light is reflected from a mineral's crystal faces or surfaces of breakage. It is closely related to refractive index, which measures how the light passes (refracts) through the mineral. As refractive index increases, the ratio of reflected light to incident light also increases. Luster is defined with terms that describe the appearance of the reflecting surface and, to varying degrees, the interior of the mineral beneath that surface. Vitreous (from the Latin word for glass) minerals are moderately reflective and have a high degree of transparency. Metallic minerals are opaque and have the appearance of polished metal. Adamantine minerals appear brilliant, like the diamond. The surface of silky minerals appears silklike due to the reflection of light from bundles of tiny parallel fibers. Waxy minerals reflect light like paraffin wax. Resinous minerals reflect like tree resin. Table A.1 shows luster in relation to other optical characteristics.

Lustrous minerals become dulled by chemical weathering and mechanical abrasion; thus, observations of luster should be made on fresh surfaces. Luster also decreases as the crystal size of the mineral decreases. The luster of pyrite changes progressively from metallic (macro-

Table A.1. Relation of Luster to Other Optical Properties

Luster	Refractive Index Range	Transparency	Mineral Examples
Subvitreous	1.3–1.5	translucent	fluorite
Vitreous	1.5–1.8	transparent to translucent	quartz, barite, calcite
Subadamantine	1.8–2.2	transparent to translucent	zircon
Adamantine	2.2–2.7	transparent to translucent	sphalerite, diamond
Submetallic	2.7–3	translucent to opaque	goethite, hematite
Metallic	>3	opaque	pyrite, chalcopyrite
Silky	variable	translucent to opaque	gypsum (satin spar); goethite
Waxy	variable	translucent	chalcedony, chert
Resinous	variable	translucent	sphalerite, garnet

crystals) to dull (microcrystalline, or sooty, pyrite), gypsum changes from vitreous (selenite) to dull (alabaster), and quartz changes from vitreous (macrocrystals) to waxy (chalcedony). Luster may be enhanced by coatings of foreign material on crystal surfaces. Examples are iridescent calcite from the coal measures and Peske Quarry and some brown fluorite from Pint's Quarry. The practiced eye can distinguish between colorless varieties of barite and calcite, because barite is somewhat more brilliant than calcite.

Hardness

Hardness is a measure of the strength of the chemical bonds in a mineral. The simplest way to determine hardness is to scratch the mineral with a reference material, the hardness of which is known. The edge or corner of the harder mineral creates a groove in the softer mineral by breaking chemical bonds along its surface. The standard reference is the Moh's scale, which consists of ten minerals arranged in order of increasing hardness from 1 to 10. If the unknown mineral scratches reference H = 5 but is scratched by reference H = 6, its hardness is between 5 and 6. In most cases, hardness may be approximated to a sufficient degree by using two reference materials—a fingernail (H = 2–3) and a knife blade (H = 5–6). A glass plate may be used in place of a knife blade. Using these references, minerals are assigned to one of three hardness categories: those softer than a fingernail, those harder than a knife blade, and those that are in between.

When making a hardness determination, please keep in mind the following. (1) If the outer part of the mineral is weathered or otherwise chemically altered, the mineral may appear to be softer than it is. For example, the chalky alteration rind surrounding some chert nodules may appear to be scratched by your fingernail, despite the fact that the hardness of chert is 7. (2) Fine-grained varieties of hard minerals like hematite, limonite, and pyrite may appear to be soft. In the latter two cases, the reason for the discrepancy in hardness is that scratching does not actually break bonds within mineral grains, rather it breaks the weak forces of cohesion between them. It may be difficult to obtain an accurate hardness for any fine-grained material. Remember that the scratch test may damage the mineral; therefore, it should be made at an inconspicuous place on the specimen.

Specific Gravity

Specific gravity is the weight of a standard volume of material compared to the weight of the same volume of water. Thus, specific gravity is a ratio of densities. Weight is typically measured in grams, and the standard volume is the cubic centimeter. Differences in specific gravity among minerals result from differences in the atomic weights of the component chemical elements and their physical arrangement (packing) in the crystal structure. Barite ($BaSO_4$, S.G. = 4.5) has a higher specific gravity than anhydrite ($CaSO_4$, S.G. = 2.9) because it contains the heavy element barium. Diamond (C, S.G. = 3.5) has a higher specific gravity than graphite (C, S.G. = 2.2) because the carbon atoms in diamond are more tightly packed than they are in graphite.

Specific gravity can be determined precisely with analytical equipment. A rough estimate can be made by hefting similar volumes of the unknown mineral and quartz (the specific gravity of quartz [2.65] is considered to be average). Metal oxides like hematite and goethite and sulfides like pyrite and galena have higher than average specific gravities. Porous materials like cellular quartz and limonite appear to have lower than average specific gravities because part of the volume is occupied by voids. Barite and calcite are often similar in appearance, but barite feels much heavier.

Cleavage and Fracture

Cleavage and fracture are terms that describe the way a mineral breaks. If the surfaces of breakage are flat, the mineral possesses cleavage; if they are not flat (even though smooth), the mineral fractures. Cleavage results from inequalities in chemical bond strength within a mineral crystal; uniform bond strength promotes fracturing.

Cleavage is generally described in terms of the number of directions of cleavage planes, the angle at which different cleavage planes intersect, and the degree of perfection of cleavage plane development. While it is possible to measure the angle between intersecting cleavage planes, it is gen-

Table A.2. Types of Cleavage and Fracture in Minerals

No. of Directions	Angular Relations	Name	Mineral Examples
1	—	basal	mica
2	perpendicular	rectangular	feldspar
2	perpendicular	prismatic	pyroxene
2	oblique	prismatic, pinacoidal	hornblende
3	perpendicular	cubic	galena
3	2 perpendicular, 1 oblique	pinacoidal	barite, anhydrite, celestite, gypsum
3	oblique	rhombohedral	calcite, dolomite
4	oblique	octahedral	fluorite
6	oblique	dodecahedral	sphalerite (rarely observed)
0	—	conchoidal fracture	quartz, chalcedony, chert

erally sufficient, for purposes of mineral identification, simply to observe whether the angle is perpendicular or oblique. Degree of perfection (expressed in terms like "perfect," "fair," and "poor") is often difficult to assess in the average mineral specimen, and it will not be considered further. Cleavage types are cataloged in table A.2 and illustrated in figure A.1.

To determine the type of cleavage, tilt the mineral back and forth and look for light reflecting from a flat surface. A cleavage surface may be uniformly flat, or it may be stair-stepped, with the light reflecting from all "treads" at the same position. Once a cleavage direction is identified, it may be confirmed by turning the mineral over and looking for a parallel reflecting surface on the opposite side. The mineral may then be viewed at different angles to determine if additional cleavages are present. Where two cleavages intersect, the angle, whether perpendicular or oblique, can be noted. The process may be repeated until all cleavage directions and their angular relations have been discovered.

Surfaces of breakage that are not flat are fracture surfaces. Many minerals cleave in some directions and fracture in others. A good example is hornblende, which fractures in a direction transverse to its two well-developed cleavages. If observation of the mineral fails to reveal any flat reflecting surfaces, then it possesses fracture only. A sizeable list of terms used to describe the nature of the fracture surface has evolved over the years. Most of these terms are unnecessary for mineral identification. Conchoidal fracture is distinctive and deserves special mention. The term means shell-like and is exemplified by the chipped edge of a piece of glass (fig. A.2). It is present in hard vitreous minerals that lack cleavage, like

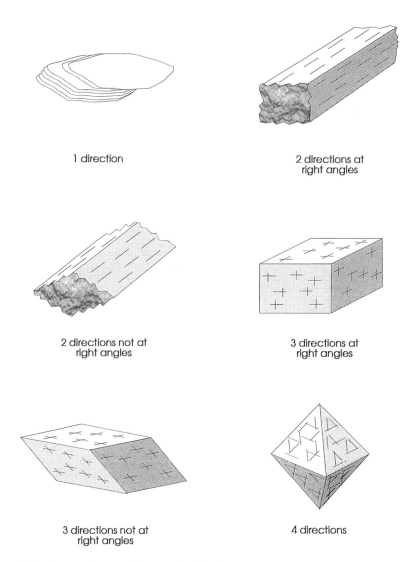

1 direction

2 directions at
right angles

2 directions not at
right angles

3 directions at
right angles

3 directions not at
right angles

4 directions

A.1. Common types of cleavage in minerals.

quartz and some garnet. It is also present in aggregates of fine mineral grains, as in chert, chalcedony, and calcite in lithographic limestone.

The process for determining cleavage works well if the candidate is a fragment of a single crystal. Where several crystals have grown together in random orientation, determining the number of cleavages and their angular relations may be difficult if not impossible. Tilting the sample back and forth in the light may reveal a confusing display of reflections at many different angles. A good example is sphalerite, which possesses six directions of cleavage but rarely occurs in isolated crystals that are large enough to permit a reliable determination. In such cases, it is best to note that the mineral possesses some cleavage and then to rely on other properties to confirm its identity.

Another problem in determining cleavage in minerals owes to the fact that the exterior growth surfaces (faces) of a mineral generally are also flat. With experience, crystal and cleavage faces may be distinguished through observation of subtle surface details, such as chemical etching and growth lines (striations). When in doubt, knowing that the mineral in question does or does not possess cleavage will help. For example, pyrite characteristically forms good crystals, but it does not have cleavage. Any reflecting flat surfaces observed in pyrite must necessarily be crystal faces.

Crystal Forms and Habits

Minerals may be identified by their crystal forms or habits. Principles and terminology used in defining forms and habits are defined in Appendix B.

Reaction to Hydrochloric Acid

Most minerals are soluble, at least to some degree, in acids. Table 2.1 reports solubility of Iowa's minerals in common acids, such as hydrochloric acid (HCl), sulfuric acid (H_2SO_4), nitric acid (HNO_3), and hydrofluoric acid (HF). Dilute HCl (5 to 10 percent) is a very useful tool for distinguishing between calcite and dolomite, two minerals that are very common in Iowa. These minerals have many physical properties in common, including hardness and rhombohedral cleavage, and it is often difficult to tell them apart. If a small drop of cold HCl is placed upon calcite it will effervesce vigorously, due to the rapid release of carbon dioxide gas. Aragonite, the

A.2. Obsidian, showing conchoidal fracture. Specimen is 8 centimeters across.

Table A.3. Key to Identification of Iowa's Minerals

Mineral Name	Chemical Composition	Color	Streak	Hardness	Specific Gravity	Cleavage/ Fracture	Crystal Form/ Habit	Miscellaneous
Anglesite	$PbSO_4$	white, gray, tan	white	3	6.3	prismatic	microcrystalline	typically occurs as coatings on galena
Anhydrite	$CaSO_4$	white, gray	white	3–3.5	2.92	prismatic	microcrystalline	
Aragonite	$CaCO_3$	white, gray	white	3.5–4	2.95	prismatic	long prismatic, acicular	
Barite	$BaSO_4$	colorless, white, yellow	white	3–3.5	4.5	prismatic	bladed, tabular, plumose, rosettes, microcrystalline	multicrystalline varieties fluoresce in long-wave UV
Biotite	$K(Mg,Fe,)_3 (Al Si_3O_{10})(OH)_2$	black, brown	white to pale brown	2–2.5	3.0	basal	flaky	altered biotite color changes to bronze
Calcite	$CaCO_3$	colorless, white, yellow, pink, brown, gray	white	3	2.71	rhombohedral	scalenohedral, rhombohedral, sparry, microcrystalline	basal and rhombohedral twins occur locally; zoning common; brown variety may exhibit purple iridescent tarnish, some varieties fluoresce in long-wave UV
Celestite	$SrSO_4$	colorless, pale blue	white	3–3.5	3.96	prismatic	bladed, tabular	
Cerussite	$PbCO_3$	colorless, gray	white	3–3.5	6.55	prismatic	microcrystalline, short prismatic	commonly occurs as crystals perched on galena

Mineral	Formula	Color	Streak	Hardness	Specific gravity	Cleavage	Habit	Remarks
Chalcedony	SiO_2	white, gray; impurities may add color, e.g., red brown, black	colorless	7	2.65	conchoidal fracture	microfibrous, botryoidal, geodal, fine concentrically banded	may exhibit iridescent tarnish
Chalcopyrite	$CuFeS_2$	metallic yellow	black	3.5–4	4.2	fracture	pseudotetrahedral, microcrystalline	
Chert	SiO_2	white, gray; impurities may add color, e.g., tan, red brown, rust brown	colorless, pale red or brown	7	2.65	conchoidal fracture	microgranular, botryoidal, nodular, coarse concentrically banded	nodules may exhibit white chalky rind due to alteration
Chlorite	complex Mg, Fe, Al, silicate	dark green to greenish gray	pale green	2–2.5	2.6–3.3	basal	microcrystalline	
Copper	Cu	metallic orange pink	black	2.5–3	8.9	none	massive	coated with brown or green oxidation product
Dolomite	$CaMg(CO_3)_2$	white, tan, pink	white	3.5–4	2.85	rhombohedral	rhombohedral, microcrystalline	rhombs often curved
Epidote	complex Ca, Fe, Al silicate	olive green to yellow green	white	6–7	3.4	prismatic	microcrystalline	occurs as veins in granite and as vesicle fillings in basalt
Fluorite	CaF_2	brown, purple, yellow	white	4	3.18	cubic	cubic	brown and yellow varieties fluoresce in long-wave UV
Galena	PbS	metallic gray	galena	2.5	7.5	cubic	cubic, sparry	
Garnet	$Fe_3Al_2(SiO_4)_3$	red brown	colorless	6.5–7.5	3.5–4.3	fracture	cubic dodecahedral	
Goethite	$FeO.OH$	dark brown	dark brown	5–5.5	4.37	fracture	fibrous, botryoidal, stalactitic	

Continued

Table A.3. *Continued*

Mineral Name	Chemical Composition	Color	Streak	Hardness	Specific Gravity	Cleavage/ Fracture	Crystal Form/ Habit	Miscellaneous
Gypsum	$CaSO_4 \cdot 2H_2O$	colorless, white, gray	white	2	2.32	prismatic	bladed, tabular, fibrous, microcrystalline	microcrystalline variety banded, larger crystals locally twinned (swallow-tail)
Hematite	Fe_2O_3	red to red brown; coarser crystals are metallic gray	red, red brown	5.5–6.5	5.26	fracture	microcrystalline	
Hemimorphite	$Zn_4(Si_2O_7)(OH_2) \cdot H_2O$	white, gray, tan	white	4.5–5	3.4	prismatic	microcrystalline	
Hornblende	complex Ca, Mg, Fe, Al silicate	dark green to black	pale green	5–6	3.0–3.4	prismatic	long prismatic, microcrystalline	
Kaolinite	$Al_2(SiO_5)(OH)_4$	white, tan	white	2	2.6	basal	microcrystalline, earthy	
Limonite	mixture of Fe oxides and hydroxides	yellow brown	rust brown	5–5.5	3.6–4.9	fracture	microcrystalline, earthy	
Malachite	$CuCO_3(OH)_2$	green to greenish black	green	3.5–4	3.9	fracture	microcrystalline	
Marcasite	FeS_2	metallic greenish gray	black	6–6.5	4.89	fracture	bladed, wedge-shaped, microcrystalline	
Millerite	NiS	metallic bronze	black	3–3.5	5.5	fracture	capillary	
Muscovite	$KAl_2(AlSi_3O_{10})(OH)_2$	colorless, silver, pale green, or pale brown	white	2–2.5	2.76–2.88	basal	flaky	

Mineral	Composition	Color	Streak	Hardness	Specific gravity	Cleavage/fracture	Crystal habit	Remarks
Plagioclase feldspar	$NaAlSi_3O_8$–$CaAl_2Si_2O_8$	white, gray	white	6	2.62–2.76	rectangular	short–long prismatic, microcrystalline	
Potassium feldspar	$KAlSi_3O_8$	pink, red, tan, white	white	6	2.5	rectangular	short prismatic, microcrystalline	
Pyrite	FeS_2	metallic brass yellow	black	6–6.5	5.02	fracture	cubic, octahedral, pyritohedral, microcrystalline	may exhibit iridescent tarnish
Pyroxene	complex Ca, Mg, Fe silicate	dark green to black	pale green	5–6	3.4–3.7	prismatic	short prismatic, microcrystalline	
Quartz	SiO_2	colorless, white, gray	colorless	7	2.65	conchoidal fracture	prismatic, drusy, massive	locally stained by limonite or hematite
Siderite	$FeCO_3$	tan, brown, red brown	white	3.5–4	3.96	rhombohedral	sparry, rhombohedral	
Smithsonite	$ZnCO_3$	white, tan, gray	white	4–4.5	4.4	rhombohedral (rarely observed)	microcrystalline	
Sphalerite	ZnS	yellow brown to dark brown	pale yellow brown	3.5–4	4.0	dodecahedral (rarely observed)	sparry, tetrahedral	crystals generally malformed and etched
Tourmaline	complex Na, Ca, Fe, Al, B silicate	black	colorless	7–7.5	3.1	fracture	long prismatic	

Source: Numerical data from Klein and Hurlbut (1993)

chemistry of which is identical to calcite's, reacts in a similar manner. When HCl is placed on coarse fragments of dolomite, it effervesces sluggishly, if at all. Some effervescence occurs when dolomite is powdered. The weathered surfaces of dolomite, or any other mineral, may contain water-deposited calcite or aragonite. Therefore, it is important that the test be performed on a fresh surface of the mineral and away from cracks.

Identification Guide

Table A.3 lists the important physical properties of the minerals that have been reported in Iowa.

Appendix B

Crystal Forms and Habits Commonly
Observed in Iowa's Minerals

Forms

Many of Iowa's mineral collectibles exhibit well-developed crystals, in part because the minerals form in cavities where growth is relatively free from interference. The shapes of individual crystals reflect the geometry of the internal crystal lattice. Lattice geometry, in turn, results from the nature and spatial distribution of the chemical bonds that hold the lattice together.

All minerals belong to one of six crystal systems, which are defined by the lengths of the crystallographic axes and the angles between them. These axes are designated a, b, and c. For purposes of graphic illustration, the c axis is usually vertical, and the a and b axes are horizontal. The systems are listed in table B.1.

Each crystal system is represented by a number of crystal forms, which are defined by the geometric arrangement of crystal faces around the crystallographic axes (fig. B.1). Some crystal forms, like cubes, are closed, meaning that the crystal faces of the form completely enclose the mineral. Other forms, like prisms, are open, meaning that two or more forms must combine in order to enclose the crystal completely. Crystal forms must be compatible with the symmetry of the crystal system.

It is common in natural crystals, even in those exhibiting closed crystal forms, for forms to be present in combination. For example, in pyrite, a cube may combine with an octahedron, resulting in a cubo-octahedron (fig. B.2e). In barite, prismatic and pinacoidal forms combine (fig. B.2p). In calcite rhombohedra, scalenohedra and prisms may combine to produce a very complex-appearing crystal (fig. B.2l). Other forms in which major minerals in Iowa occur are illustrated in figure B.1a–t.

Habits

The crystal habit of a mineral is its common mode of appearance. Crystal habit may be defined by a specific crystal form or combination of forms, as

Table B.1. The Six Crystal Systems

Crystal System	Relative Lengths of Crystal Axes	Angles between Crystal Axes
Triclinic	$a \neq b \neq c$	$\alpha \neq \beta \neq \gamma \neq 90°$
Monoclinic	$a \neq b \neq c$	$\alpha = \gamma = 90° \neq \beta$
Orthorhombic	$a \neq b \neq c$	$\alpha = \beta = \gamma = 90°$
Hexagonal	$a = b \neq c$	$\alpha = \beta = 90° \quad \gamma = 120°$
Tetragonal	$a = b \neq c$	$\alpha = \beta = \gamma = 90°$
Isometric	$a = b = c$	$\alpha = \beta = \gamma = 90°$

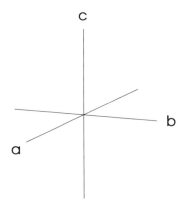

B.1. The three crystallographic axes.

discussed previously. As examples, calcite from the Cedar Rapids Quarry is scalenohedral (fig. B.2g); pyrite from Pint's Quarry is cubo-octahedral (fig. B.2e). Habit may be defined by the general shape of a single crystal. Crystals for which the c axis is longer than the a and b axes are referred to as prismatic or columnar. Gypsum and barite from several localities in Iowa display prismatic habit (B.2q, s). If c is much shorter than a and b, the habit is tabular, platy, or micaceous. Some barite from the Linwood Mine is tabular (fig. B.2p). Crystals that are needle-shaped are acicular or capillary, as exemplified by millerite from Conklin Quarry. Crystals with unequal a, b, and c axes are referred to as bladed (e.g., barite from the Peske and Pint's quarries; marcasite from the Moscow Quarry).

Habit may also be defined by the overall shape resulting from an aggregation of many individual crystals. A mineral consisting of small crystals lining a cavity is said to be drusy (e.g., quartz in some geodes). A bundle of parallel acicular crystals is called fibrous (e.g., gypsum veins from Fort Dodge). A mass of acicular crystals arranged like the spokes of a wheel, or the elements of an oriental fan, is referred to as radiating or stellate (e.g., millerite from Conklin Quarry). Banded habit describes minerals in

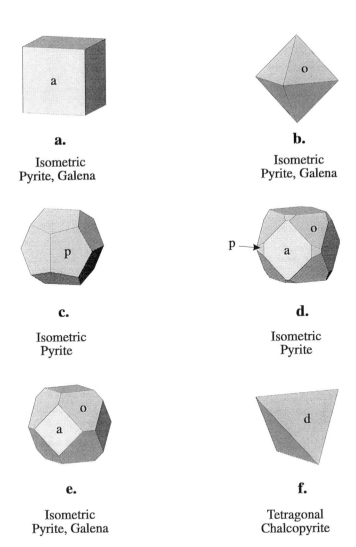

a.

Isometric
Pyrite, Galena

b.

Isometric
Pyrite, Galena

c.

Isometric
Pyrite

d.

Isometric
Pyrite

e.

Isometric
Pyrite, Galena

f.

Tetragonal
Chalcopyrite

B.2. Ideal crystal forms commonly observed in some Iowa minerals. For each crystal drawing, the system to which the crystal form or forms belong and a mineral example are given. a. cube (a); b. octahedron (o); c. pyritohedron (p); d. cube (a), modified by octahedron (o) and pyritohedron (p); e. cube (a), modified by octahedron (o); f. tetragonal disphenoid (pseudotetrahedron) (d); g. acute scalenohedron (s); h. acute rhombohedron (r); i. obtuse rhombohedron (r); j. obtuse rhombohedron (r), modified by acute scalenohedron (s); k. acute scalenohedron (s), modified by obtuse rhombohedron (r); l. prism (m), modified by acute scalenohedron (s) and obtuse rhombohedron (r); m. scalenohedron basal twin; n. rhombohedral twin; o. prism (m), modified by positive and negative rhombohedra (r+ and r−); p. basal pinacoid (c), vertical rhombic prism (m_v), and horizontal rhombic prism (m_h); q. vertical rhombic prism (m_v), with two horizontal rhombic prisms (m_{h1} and m_{h2}); r. horizontal rhombic prism (m_h), with two vertical rhombic prisms (m_{v1} and m_{v2}); s. side pinacoid (b), with vertical rhombic prism (m_v) and inclined rhombic prism (m_i); t. "swallow-tail" twin.

g.

Trigonal
Calcite

h.

Trigonal
Calcite

i.

Trigonal
Calcite, Dolomite

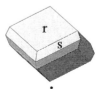

j.

Trigonal
Calcite

k.

Trigonal
Calcite

l.

Trigonal
Calcite

m.

Trigonal
Calcite Basal Twin

n.

Trigonal
Calcite Rhombohedral Twin

B.2. *Continued*

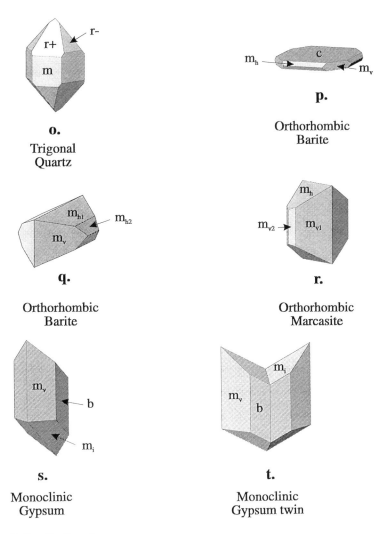

o.
Trigonal
Quartz

p.
Orthorhombic
Barite

q.
Orthorhombic
Barite

r.
Orthorhombic
Marcasite

s.
Monoclinic
Gypsum

t.
Monoclinic
Gypsum twin

B.2. *Continued*

thin layers defined by differences in color or texture (e.g., bedded gypsum from Fort Dodge and Sperry gypsum deposits and flowstone in some limestone caves) (fig. 3.14). Stalactitic and stalagmitic are terms used to describe cylindrical masses of minerals, the axes of which are all parallel to each other (e.g., calcite in limestone caves, and pyrite at the Linwood Mine and the Lafarge Quarry). Colloform minerals are aggregated in spheroidal groups, like a cluster of grapes or the head of a cauliflower. In cross section, colloform masses are typically radiating acicular (e.g., pyrite from Moscow Quarry and some chalcedony in geodes) (fig. 3.26). Dendritic means having a branching treelike or fernlike appearance (e.g., manganese oxide minerals deposited upon a fracture surface in a rock; manganese dendrites are commonly misidentified as plant fossils). Plumose means feathery in appearance (e.g., barite from the Linwood Mine). Sparry refers to coarse masses exhibiting good cleavage (e.g., calcite and sphalerite from numerous localities). Spheroidal means spherical in shape (e.g., geode exteriors and chert concretions from the Cedar Rapids Quarry) (fig. 3.8). Nodular refers to smooth-surfaced knobby masses (e.g., chert from numerous localities). Microcrystalline means that the individual crystals are too small to be visible without a microscope. Iron sulfide inclusions in calcite from many localities are microcrystalline.

The crystal illustrations in the color section, like those in textbooks and reference books on crystallography, represent ideal crystals, that is, crystals that are perfectly proportioned. In nature, ideal crystal growth rarely, if ever, is achieved. Lack of adequate growth space, rapid growth rate, and interruptions during growth, among other factors, may cause crystals to become malformed. Crystals growing from a base lack lower terminations. Crystals growing in a narrow crack tend to become flattened. In such cases, identifying the crystal forms may be very difficult unless the identity of the mineral is known.

Appendix C

Institutions That House Mineral Collections
and Rock- and Mineral-Collecting Organizations
in Iowa

Institutions

Anderson Museum
Cornell College*
Mount Vernon, IA 52314
(319) 895-4000
http://www.cornell-iowa.edu

Department of Earth Science
121 Latham Hall
University of Northern Iowa*
Cedar Falls, IA 50613
(319) 273-2759
http://www.uni.edu/museum

Department of Geography
   and Geology
Drake University*
Des Moines, IA 50311
(515) 271-2967
http://www.geo.drake.edu

Department of Geological and
   Atmospheric Sciences
Iowa State University*
253 Science I
Ames, IA 50010
(515) 294-4477
http://www.geology.iastate.edu

Department of Geology
University of Iowa*
121 Trowbridge Hall
Iowa City, IA 52242
(319) 335-1818
http://www.geology.uiowa.edu

Geological Survey Bureau*
Iowa Department of Natural
   Resources
109 Trowbridge Hall
Iowa City, IA 52242
(319) 335-1575
http://samuel.igsb.uiowa.edu

Grout Museum of History
   and Science
503 South Street
Waterloo, IA 50701
(319) 234-6357
http://iowa-counties.com/
   blackhawk/grout-museum/
   grout.htm

Museum of Natural History
University of Iowa*
Macbride Hall
Iowa City, IA 52242
(319) 335-0481
http://cgrer.uiowa.edu/
   iowa-environment.museum/
   Museum/html

Porter House Museum
401 West Broadway
Decorah, IA 52101
(319) 382-8465

Putnam Museum
1717 West 12th Street
Davenport, IA 52804
(319) 324-1933
http://www.qonline.com/arts/
    putnam/index.html

Sanford Museum and Planetarium
117 East Willow
Cherokee, IA 51012
(712) 225-3922

State Historical Society of Iowa
East 6th and Locust Streets
Des Moines, IA 50309
(515) 281-5111
http://www.uiowa.edu./~shsi/
    loc/locdm/htm

Organizations*

Ames Rock and Mineral Club
Ames

Blackhawk Gem and Mineral Society
Cedar Falls

Cedar Valley Rock and Mineral
    Society
Cedar Rapids

Central Iowa Mineral Society
Norwalk

Chicauqua Rockhound Society
Mount Pleasant

Dallas County Rock Club
Des Moines

Des Moines Lapidary Society
Des Moines

Des Moines Valley Rocks and Relics
    Club
Farmington

Geode Rock and Mineral Society
New London

Nishna Valley Rock Club
Atlantic

River Valley Rockhound Club
Fort Dodge

Sac and Fox Lapidary Club
Ottumwa

Siouxland Gem and Mineral Society
Sioux City

Storm Lake Area Rockhounds
Storm Lake

---

*Current addresses may be obtained from the Midwest Federation of Mineralogical
and Geological Societies by contacting the secretary, Mary Hanning, at Route 1, Box
63, Huntsville, IL 62344, phone (217) 667-2285.

Glossary

acicular referring to crystals that are needle-shaped; example: millerite.

acid mine drainage waters issuing from surface and underground mines that are high in acids and dissolved metals.

acute referring to crystals, the faces of which meet in acute angles (see Appendix B, fig. B.2b, g).

alluvial pertaining to rivers and streams.

alteration physical and/or chemical changes in rocks and minerals.

anhedral referring to minerals that fail to develop well-formed crystals.

anion a negatively charged ion.

aqueous pertaining to water.

assemblage a group of rocks or minerals having a common origin.

attenuated referring to crystals that are elongated in a particular direction.

basalt a dark-colored, fine-grained igneous rock containing plagioclase feldspar, pyroxene, and miscellaneous other silicate minerals.

basal twin a crystal twin, in which the common surface between the two individual parts is parallel to the base of the crystal (see Appendix B, fig. B.2m); example: calcite.

basin a large, generally water-filled depression into which sediments are deposited.

bedding plane a surface, usually horizontal, that separates layers of sediment or sedimentary rock.

bedrock a continuous solid rock mass.

biogenic having a biological origin.

bituminous coal coal intermediate between lignite and anthracite; "soft" coal.

bladed referring to crystals that are shaped like the blade of a knife, i.e., the three crystallographic axes are of different lengths; examples: barite, gypsum, marcasite.

bleb a tiny, spherically shaped mass.

brachiopod a bivalved marine invertebrate.

breakdown rocks on the floor of a cave that were detached from its ceiling or walls.

breccia sedimentary rock composed of angular fragments.

brine water with abnormally high salt content.

calcareous containing calcium carbonate.

calcine to drive water from gypsum by heating.

capillary referring to crystals that are shaped like a human hair; example: millerite.

carbonate rock sedimentary rock composed of calcite or dolomite.

cation a positively charged ion.

cavern a large opening created by dissolution of carbonate rock.

cavity any opening created by dissolution of carbonate rock.

cementation a process by which loose sediment becomes sedimentary rock.

chemical weathering disintegration of minerals and rocks at the surface by breaking chemical bonds.

chert microgranular form of quartz.

cleat fracture in coal that is transverse to bedding.

cleavage planar break surface in a mineral.

coal ball concretionary mass of siltstone occurring in coal beds.

coalification process by which decaying organic matter becomes coal.

collapse breccia breccia formed through collapse of a cave roof.

colloform referring to an aggregate of crystals, the overall shape of which is rounded or globular; example: chalcedony.

compaction pressure-induced consolidation of fine-grained sediment.

concentric banding banding consisting of a series of nested spheres.

concretion spheroidal mass of mineral matter formed by replacement of sedimentary rock; commonly chert or pyrite.

conodont tiny toothlike fossils of uncertain origin.

crevice a dissolution-enlarged vertical fracture.

crinoid marine invertebrate animal consisting of a central body containing branching arms and fastened upon a flexible stalk.

crust a thick coating of mineral matter.

crystal face plane-shaped growth surface on a mineral.

crystalline consisting of atoms, ions, or molecules joined together in a regular symmetrical array.

cubo-octahedron a crystal form in which the eight corners of the cube are beveled by the faces of an octahedron (see Appendix B, fig. B.2e); examples: pyrite, galena.

damp vapors given off when coal is exposed to the atmosphere; may be toxic when allowed to accumulate in poorly ventilated underground mines.

density the mass (generally in grams) of a standard volume (generally in cubic centimeters) of material.

dessication process (usually evaporative) by which water is removed from sediment or rock.

diagenesis chemical and/or physical changes occurring in a sedimentary rock after deposition of sediment and prior to metamorphism.

disphenoid a crystal form in which the faces of the upper half consist of two isosceles triangles joined at the base and which alternate with two joined isosceles triangles in the lower half (see Appendix B, fig. B.2f); example: chalcopyrite.

dodecahedron a twelve-sided crystal, each face of which is either rhombic or pentagonal in shape (see Appendix B, fig. B.2c); example: garnet.

dolomitization process by which limestone is chemically changed to dolostone.

dolostone a carbonate rock consisting chiefly of the mineral dolomite.

doubly terminated referring to a crystal in which the crystal faces are well developed at both ends.

dragline an excavating machine having a bucket that is dropped from a boom and dragged toward the machine with a cable.

drift mining underground mining in which the workings are at about the same level as the entrance.

drill core a cylindrical mass of rock obtained during drilling.

dripstone mineral formed by the dripping action of water in a cave.

drusy referring to an aggregate of small crystals coating a rock surface; example: quartz in some geodes.

effervescence bubbling action caused by rapid release of carbon dioxide gas from carbonate minerals.

efflorescence whitish powder encrusting a rock surface due to evaporation of water.

electron microscope optical system in which the energy source is a focused beam of electrons; capable of magnifying thousands of times actual size.

embayment a deep indentation or recess of a shoreline.

encrustation a hard coating of minerals on a rock surface or the process of forming the coating; var. *incrustation*.

epicontinental pertaining to the continental shelf or interior.

epigenetic referring to minerals that form after the formation of the enclosing rock.

epithermal referring to a hydrothermal mineral deposit formed under shallow, low-temperature conditions.

erosion mechanical destruction and removal of loose material by running water, wind, or glacial ice.

erratic (glacial) a rock fragment carried by a glacier and deposited at some distance from its source.

esker a long sinuous ridge of sand and gravel deposited by a stream flowing beneath or within a glacier.

etch to pit or otherwise roughen the surface of a mineral by chemical dissolution.

euhedral referring to minerals that develop well-formed crystals.

evaporite a mineral formed by precipitation from hypersaline water; examples: halite, gypsum.

exoskeleton the external skeleton of an animal; example: a clam shell.

fire clay a clay formed of hydrous silicate minerals that is capable of withstanding high temperature without destruction.

fissure a rock fracture along which there is distinct separation.

flat referring to horizontal, sheetlike bodies of ore mineral.

float a rock displaced from its bedrock source.

flocculent referring to an aggregate of tiny crystals that are loosely held together in clumps.

fluvial stream related.

formation a fundamental part of the rock record consisting of a body of rock that possesses characteristics, such as composition, texture, or fossils, that distinguish it from the surrounding rocks.

friable easily crumbled.

fusilinid a wheat grain–shaped marine invertebrate animal.

fusion crust blackened crust on the surface of a meteorite caused by frictional heating during passage through the Earth's atmosphere.

Gabion basket a wire mesh enclosure containing crushed rock or gravel used to control erosion on steep slopes.

gad a pointed tool used to remove a mineral specimen from a cavity.

gash vein an ore-filled, dissolution-enlarged vertical fracture.

geodal pertaining to a geode.

geode a crystal-lined spheroidal cavity that is separable from its enclosing rock.

gin a machine used for hoisting rock from a mine.

glacio-fluvial pertaining to glacial and stream activity.

granite a coarse-grained igneous rock consisting primarily of quartz and feldspar.

granule a rounded mineral grain, in size like coarse sand.

gravity study study involving the measurement of differences in gravitational acceleration.

hardness a measure of the strength of the chemical bonds in a mineral; usually determined by scratching an unknown with a reference material of known hardness.

heavy mineral a mineral the specific gravity of which is greater than that of quartz.

helictitic like a stalactite or stalagmite but with a twisted shape.

high wall vertical face in an open-cut mine or quarry.

hydration process by which a rock or mineral absorbs water.

hydrothermal pertaining to mineral deposits that form through the action of heated subsurface water.

hypersaline referring to water having much greater than normal dissolved salt content.

igneous referring to rocks and minerals formed by the solidification of molten silicate liquid and to the process of formation.

inclusion a small crystal of a mineral wholly contained within a different mineral; example: pyrite inclusions in calcite.

indurated hardened, as sediment, by compaction, cementation, or heat.

interstitial referring to the voids between mineral grains.

intrusive referring to the igneous processes that occur below the Earth's surface.

invertebrate referring to an animal that lacks a backbone.

iridescent referring to minerals that display rainbowlike colors on their crystal surfaces or within their interiors.

iron meteorite a meteorite composed mainly of metallic iron and nickel.

isotope one of two or more species of the same chemical element that are distinguished by the number of neutrons in their nuclei.

kame a low hill composed of poorly sorted glacial material.

karst type of topography formed upon carbonate rocks; characterized by sinkholes and caves.

lamination fine-scale layering in sedimentary rock.

lapidary a cutter and polisher of colored stones.

leach to remove by chemical dissolution.

lensoid referring to a body of material that is thick in the middle and thins toward its edges.

lignite brown coal that is intermediate between peat and bituminous coal.

lithic fragment a rock fragment.

lithification the process (compaction, cementation, recrystallization) by which loose sediment becomes sedimentary rock.

loess wind-deposited glacial dust.

longwall mining coal mining method in which the entire coal seam is removed and in which the roof is supported by timbers and waste rock.

macrocrystalline referring to crystals that are large enough to be seen with the naked eye.

mafic rich in magnesium and iron minerals.

magma molten silicate liquid.

magnetic study study involving the measurement of differences in magnetic susceptibility of rocks in the Earth's crust.

malformed referring to crystals for which the individual crystals are not ideally developed, resulting in a distorted appearance.

mantle region of silicate rock in the Earth lying beneath its crust and above its core.

massive referring to a formless mass of small crystals.

mass wasting movement of soil and rock downslope without the aid of a flowing medium such as running water or glacial ice.

matted referring to a flattened mass of acicular crystals.

member subdivision of a formation.

metamorphism the action of heat and pressure on rocks below the Earth's surface, which causes physical and chemical changes in them.

meteor a physical object of extraterrestrial origin traveling through the Earth's atmosphere.

meteorite a meteor that has reached Earth without complete vaporization.

microcrystalline referring to crystals that are not large enough to be seen with the naked eye.

microtektite a microscopically-sized tektite.

mineral a naturally occurring, chemically homogeneous, crystalline solid.

mineralization the process by which minerals are formed in nature.

mollusk a marine invertebrate animal; includes gastropods, pelecypods, and cephalopods.

monocrystalline referring to minerals that appear as scattered individual crystals.

monolith a single mass of rock of large size.

monosulfide chemical compound in which sulfide ion forms simple bonds with metal cations.

mudstone name given to a sedimentary rock consisting of clay-sized particles but lacking distinctive layering.

multicrystalline referring to minerals that appear as aggregates of many individual crystals.

nodule a rounded mass of hard mineral characterized by knobby protrusions usually having a different composition than the enclosing material.

obtuse referring to crystals, the faces of which meet in obtuse angles (see Appendix B, fig. B.2h, i, j).

octahedron an eight-sided crystal, each face of which is an equilateral triangle (see Appendix B, fig. B.2b); example: pyrite.

offlap withdrawal of the sea from a given area of land surface.

onlap advance of the sea over a given area of land surface.

onyx variety of cryptocrystalline quartz (chalcedony) characterized by planar laminations of alternating light and dark color.

overbank area adjacent to an active stream channel.

overburden soil and/or rock overlying economically productive rock.

oversaturation chemical condition in which the amount of material dissolved in solution is greater than the solution can hold under equilibrium conditions.

oxidizing referring to a chemical environment characterized by the presence of abundant free oxygen.

paleokarst dissolution cavities formed in limestone, dolostone, or gypsum at some time in the distant geologic past.

paleontologist one who studies the fossil remains of ancient life.

paragenesis the sequence of events in the formation of a mineral deposit.

peat deposit of unconsolidated, incompletely decomposed plant material; the first stage in the formation of coal.

pegmatite igneous rock with a composition similar to granite; characterized by unusually large crystals and exotic mineral species.

penetration twin twin in which the individual crystals appear to penetrate each other.

phantom crystal transparent or translucent crystal within which an earlier stage of mineral growth is outlined by dust or microscopic bubbles; example: calcite.

phosphatic rich in phosphorous-containing minerals.

phreatic referring to the region of groundwater that lies beneath the water table.

physical weathering weathering process by which rocks and minerals physically disintegrate.

pit a small depression in the surface of the crystal, most often caused by chemical dissolution.

pitch   vertical or inclined sheetlike bodies of ore mineral.

placoderm   a kind of Devonian fish characterized by a bony head covering.

plumose   referring to a crystal aggregate with a plumelike or feathery shape; example: barite.

polymorph   one of two or more crystallographic forms of a given chemical substance; examples: graphite and diamond are polymorphs of carbon.

pore water   water that fills the interstices in sediments and rocks.

postformational   referring to an event or process that occurs after the formation of a rock or mineral.

Precambrian Shield   large area of exposed Precambrian rock; typically forms the nucleus of a continent.

precipitation   chemical process whereby chemicals in solution crystallize to form minerals.

prismatic   referring to a crystal that is shaped like a prism; example: barite.

protolith   a rock prior to its conversion to a metamorphic rock.

protore   unoxidized and unenriched primary sulfide deposit.

pry bar   an elongated steel bar used for extracting minerals recessed in cavities.

pseudomorph   a mineral, the outward form of which is that of a different mineral; commonly results from the replacement of macrocrystals of a mineral by microcrystals of the other mineral; example: goethite pseudomorphic after pyrite.

pyritic   containing abundant pyrite.

pyritohedron   a twelve-sided crystal, each face of which is pentagonal (see Appendix B, fig. B.2c); example: pyrite.

quarry   a surface excavation from which commercially valuable rocks are produced.

quartzite   a metamorphic rock formed from quartz sandstone.

radiating   referring to an aggregate of long prismatic, bladed, or acicular crystals that appear to have grown outward from a common center; example: millerite.

recrystallization   process by which new grains of a given mineral form in response to diagenetic or metamorphic changes in the environment.

reducing   referring to a chemical environment characterized by the lack of free oxygen.

regression   retreat of the sea from land areas.

relief   variations in height and slope of the land surface.

replacement   process by which a mineral is chemically removed and its place taken by another mineral.

residual material remaining after physical and chemical removal of mineral substance.

rhombohedron an eight-sided crystal, each face of which is a rhombus (see Appendix B, fig. B.2h, i); example: calcite.

rind the outer layer of a geode or nodule.

room and pillar mining process of the underground mining of horizontal beds (as of limestone or coal) in which all that is valuable is removed, except for regularly spaced columns that are left to support the ceiling of the mine.

rosette a cluster of tabular crystals, arranged like the petals of a rose; examples: gypsum, barite.

sandstone a sedimentary rock consisting of sand-sized grains of sediment (predominantly quartz).

scalenohedron a twelve-sided crystal, each face of which is a scalene triangle (see Appendix B, fig. B.2g); example: calcite.

scraper a type of animal- or machine-powered earthmoving equipment.

secondary mineral a mineral that forms by the chemical replacement of a preexisting mineral.

sediment loose solid material, being or having been transported by an agent such as running water or glacial ice.

sedimentary pertaining to sediment.

seismic pertaining to earthquakes or other Earth vibrations.

septarian nodule a large concretion containing radiating cracks that widen toward the center of the concretion and that are typically lined or filled with mineral substance, most often calcite.

shaft mining method of underground mining in which the productive material is reached by a vertical or steeply inclined opening.

shale sedimentary rock consisting of finely layered, compacted, clay-sized sediment.

silicate chemical compound made up of structural units containing silicon and oxygen.

siltstone sedimentary rock consisting of silt-sized particles of sediment.

sinkhole a more or less circular opening in the land surface created by the collapse of the roof of an underlying cave.

skarn a calcium silicate–bearing rock formed by the metamorphism of limestone.

slope mining a type of underground mining in which the productive material is reached by a slope that inclines at a low angle downward from the surface.

solution-enlargement process by which a fracture or other space in a rock is increased in size by chemical dissolution of the rock.

sparry referring to aggregates of crystals displaying well-developed cleavage and lustrous surfaces; examples: calcite, sphalerite.

speleothem mineral deposit precipitated by water in a cave.

spheroidal referring to aggregates of crystals, the overall shape of which approaches spherical; example: chalcedony.

spoil rock and mineral waste remaining from surface or underground mining.

stalactite cylindrical mass of mineral deposit projecting downward from the ceiling of a cave.

stalagmite cylindrical mass of mineral deposit projecting upward from the floor of a cave.

stellate referring to aggregates of crystals in a starlike arrangement; similar to radiating; example: calcite.

stony meteorite a meteorite composed of iron and magnesium silicate minerals.

stratigraphic referring to a layered sequence of rocks.

strip mining type of surface mining in which productive material is reached by removing layers of overburden; open-cut mining.

subhedral referring to minerals, the crystal faces of which are only partly developed.

subsidence localized downward settling or sinking of earth material.

subtidal at a level below low tide.

sump the lowest level in a mine or quarry, built to collect water.

supratidal at a level above high tide.

swallow-tail twin crystal twin in which the two individual parts are joined to produce a shape like arrow fletching or a swallow's tail (see Appendix B, fig. B.2t); example: gypsum.

tabular referring to crystals that are wafer- or lozenge-shaped; examples: barite, marcasite.

tabulate coral a type of colonial coral in which the individual coral organisms are separated from each other by plates.

tectonic referring to the structure and deformation of the Earth's crust.

tektite small, rounded, pitted mass of dark-colored, silicate-rich glass, resulting from meteorite impact melting of terrestrial material.

terrace a former river floodplain, elevated above the present floodplain by downcutting of the river.

tetrahedron a four-sided crystal, each face of which is an equilateral triangle.

trace element a chemical element present in minute quantities in a mineral.

transgression encroachment of the sea upon land areas.

travertine banded deposit, typically of calcite, that forms on cave floors or other cave surfaces.

tufted referring to an aggregate of acicular or capillary crystals that appear like tufts of hair; example: millerite.

twin a symmetrical intergrowth of two or more crystals of the same mineral; the relations between individual parts of the twin are constrained by laws of symmetry.

ultrasonic cleaner a device used for cleaning minerals in which high-energy sound waves in water agitate loose particles from mineral surfaces.

ultraviolet fluorescence visible light emission from minerals by irradiation with ultraviolet light.

unconsolidated referring to sediment that is not compacted or cemented.

underclay fine-grained sediment lying immediately beneath a coal bed.

undersaturation chemical condition in which the amount of material dissolved in solution is less than the material can hold under equilibrium conditions.

vadose referring to the region of groundwater that lies above the water table.

vein a sheetlike body of mineral filling a fracture in the enclosing rock.

void a space in solid rock or between particles of sediment.

vug a small cavity.

weathering chemical and physical disintegration of rock at or near the surface of the Earth.

x-ray powder diffraction method of mineral analysis in which the mineral is bombarded with a beam of x-rays for the purpose of identifying the mineral and obtaining information about its structural and chemical properties.

zone internal layering in a crystal caused by changes in conditions during growth.

Bibliography

General

Anderson, Raymond R. 1982. "Mineral Production in Iowa." *Iowa Geology* 7: 18–19.

Anderson, Wayne I. 1983. *Geology of Iowa*. Iowa State University Press, Ames.

Grant, Stanley C. 1979. "Iowa's Mineral Heritage." *Iowa Geological Survey Open File Report*, November.

Hall, James, and J. D. Whitney. 1858. *The Geology Survey of the State of Iowa*, vol. 1, part 1.

Hinrichs, Gustavus D. 1868. "Notes on Iowa Minerals." *Second Annual Report to the State Geologist*, 267–278.

Horick, Paul J. 1974. "The Minerals of Iowa." *Iowa Geological Survey Educational Series* 2.

———. 1988. "Rock and Mineral Collecting." *Iowa Geology* 13: 16–17.

Keyes, Charles R. 1893. "Annotated Catalogue of Minerals." *Iowa Geological Survey Annual Report* 1: 181–196.

———. 1893. "Geology of Lee County." *Iowa Geological Survey Annual Report* 3: 305–408.

———. 1893. "Geology of Des Moines County." *Iowa Geological Survey Annual Report* 3: 409–492.

———. 1893. "Iowa Mineralogical Notes." *Proceedings of the Iowa Academy of Science* 3: 19–22.

———. 1896. "Geology of Johnson County." *Iowa Geological Survey Annual Report* 5: 33–116.

Klein, Cornelis, and Cornelius S. Hurlbut Jr. 1993. *Manual of Mineralogy*, 21 ed. John Wiley and Sons, New York.

Lemish, John. 1969. *Mineral Deposits of Iowa*. Iowa Southern Utilities Company, Centerville.

Prior, Jean C. 1994. "Minerals." *Iowa Geology* 19: 14–19.

Rose, J. N. 1967. "Fossils and Rocks of Eastern Iowa." *Iowa Geological Survey Educational Series* 1: 147.

1. Origins of Iowa's Minerals

Anderson, Wayne I., and Paul L. Garvin. 1984. "The Cedar Valley Formation (Devonian), Black Hawk and Buchanan Counties: Carbonate Facies and Mineralization." *Geological Society of Iowa Guidebook* 42.

Bard, Gary G. 1982. "Petrology and Diagenetic Features of the Fort Dodge Gypsum Beds." Ph.D. dissertation, Iowa State University, Ames.

Benner, Jacob A., Paul L. Garvin, and Raymond R. Rogers. 1997. "Sedimentology and Taphonomy of Pennsylvanian Paleokarst Deposits in the Middle Devonian Davenport Member, Wapsipinicon Formation, Scott County, Iowa." Abstracts with Programs. Geological Society of America, North Central Section, Madison, Wisconsin.

Blatt, Harvey, Gerard Middleton, and Raymond Murray. 1980. *Origin of Sedimentary Rocks*, 2d. ed. Prentice-Hall, Englewood Cliffs, New Jersey.

Bleifuss, Rodney L. 1972. "The Iron Ores of Southeastern Minnesota." In *Geology of Minnesota*, ed. P. K. Sims and G. B. Morey. Minnesota Geological Survey Centennial Volume, 498–505. St. Paul, Minnesota.

Brannon, Joyce C., F. C. Podosek, and Roger K. McLimans. 1992. "Alleghenian Age of the Upper Mississippi Valley Zinc Lead Deposit Determined by Rb-Sr Dating of Sphalerite." *Nature* 356: 509–511.

Brown, C. Ervin. 1967. "Fluorite in Crystal-lined Vugs in the Maquoketa Shale at Volga, Clayton County, Iowa." *American Mineralogist* 52: 1735–1750.

Chowns, T. M., and J. E. Elkins. 1974. "The Origin of Quartz Geodes and Cauliflower Cherts through Silicification of Anhydrite Nodules." *Journal of Sedimentary Petrology* 44: 885–903.

Cody, Robert D., Raymond R. Anderson, and Robert M. McKay. 1996. *Geology of the Fort Dodge Formation (Upper Jurassic), Webster County, Iowa.* Geological Survey Bureau, Iowa Department of Natural Resources Guidebook Series No. 19, Iowa City.

Craig, James R., David J. Vaughan, and Brian J. Skinner. 1988. *Resources of the Earth.* Prentice-Hall, Englewood Cliffs, New Jersey.

Dasenbrock, Frederick E. 1984. "Petrology and Diagenesis of the Fort Dodge Gypsum in the North Welles Quarry." Master's thesis, Iowa State University, Ames.

Dorheim, F., D. Koch, M. Parker, B. Peterson, J. Prior, R. Runvik, and L. Sendlein. 1972. "Tour of the U.S. Gypsum Company's Sperry Mine and Plant." *Eighth Forum on the Geology of Industrial Minerals, Field Trip Guidebook* 10–27.

Garvin, Paul L. 1982. "Sulfide Mineralization at Mineral Creek Mines, Allamakee County, Iowa." *Proceedings of the Iowa Academy of Science* 89: 44–49.

———. 1984a. "Hydrothermal Mineralization of the Mississippi-Valley Type at the Martin-Marietta Quarry, Linn County, Iowa." *Proceedings of the Iowa Academy of Science* 9: 70–75.

———. 1984b. "Mineralization at Conklin Quarry." In *Underburden/Overburden: An Examination of Paleozoic and Quaternary Strata at the Conklin*

Quarry Near Iowa City, ed. B. Bunker and G. Hallberg. Geological Society of Iowa Guidebook 41.

———. 1995. "Paleokarst and Associated Mineralization at the Linwood Mine, Scott County, Iowa." *Journal of the Iowa Academy of Science* 102: 1–21.

———, and David M. Crawford. 1992. "The Minerals of the Linwood Mine, Scott County, Iowa." *Mineralogical Record* 23: 231–238.

———, and Gregory A. Ludvigson. 1988. "Mineralogy, Paragenesis and Stable Isotopic Compositions of Mineral Deposits Associated with Late Paleozoic Karst Fills in Johnson County, Iowa." In *New Perspectives on the Paleozoic History of the Mississippi Valley: An Examination of the Plum River Fault Zone*, ed. G. Ludvigson and B. Bunker. Great Lakes Section, Society of Economic Paleontologists and Mineralogists, 18th Annual Field Conference Guidebook, Iowa City.

———, and ———. 1993. "Epigenetic Sulfide Mineralization Associated with Pennsylvanian Paleokarst in Eastern Iowa." *Chemical Geology* 105: 271–290.

———, ———, and Edward M. Ripley. 1987. "Sulfur Isotope Reconnaissance of Minor Metal Sulfide Deposits Fringing the Upper Mississippi Valley Zinc-Lead District."*Economic Geology* 82: 1386–1394.

Grant, Stanley C. 1979. "Iowa's Mineral Heritage." *Iowa Geological Survey Open File Report*, November.

Hatch, Joseph R., Matthew Avcin, W. Wedge, and L. Grady. 1976. "Sphalerites in Coals from Southeastern Iowa, Missouri, and Southeastern Kansas." *United States Geological Survey Open File Report*, 76–96.

Hayes, David T. 1986. "Origin of Fracture Patterns and Insoluble Minerals in the Fort Dodge Gypsum, Webster County, Iowa." Master's thesis, Iowa State University, Ames.

Hayes, John B. 1964. "Geodes and Concretions from the Mississippian Warsaw Formation, Keokuk Region, Iowa, Illinois, Missouri." *Journal of Sedimentary Petrology* 34: 123–133.

Heckel, Philip A. 1986. "Sea-level Curve for Pennsylvanian Eustatic Marine Transgressive-Regressive Depositional Cycles along Midcontinent Outcrop Belt." *Geology* 14: 330–334.

Heyl, Allen V., Allen F. Agnew, Erwin J. Lyons, and Charles H. Behre Jr. 1959. "The Geology of the Upper Mississippi Valley Zinc-Lead District." *United States Geological Survey Professional Paper* 309.

Hinds, Henry. 1909. "Coal Deposits in Iowa." *Iowa Geological Survey Annual Report* 19: 21–396.

Hinrichs, Gustavus D. 1868. "Notes on Iowa Minerals." *Second Annual Report to the State Geologist*, 267–278.

———. 1905. *The Amana Meteorites*. Carl Gustavus Hinrichs, St. Louis.

Horick, Paul J. 1974. "The Minerals of Iowa." *Iowa Geological Survey Educational Series* 2.

Howell, Jesse V. 1915. "The Iron Ore Deposits Near Waukon, Iowa." *Iowa Geological Survey Annual Report* 25: 33–102.

Kutz, Keith B., and Paul G. Spry. 1989. "The Genetic Relationship between

Upper Mississippi Valley District Lead-Zinc Mineralization and Minor Base Metal Mineralization in Iowa, Wisconsin, and Illinois." *Economic Geology* 84: 2139–2154.

Maliva, Robert G. 1987. "Quartz Geodes: Early Diagenetic Silicified Anhydrite Nodules Related to Dolomitization." *Journal of Sedimentary Petrology* 57: 1054–1059.

McLimans, Roger K. 1977. "Geological, Fluid Inclusion, and Stable Isotope Studies of the Upper Mississippi Valley Zinc-Lead District, Southwest Wisconsin." Ph.D. dissertation, Pennsylvania State University, University Park.

Muilenburg, Garrett A. 1914. "On the Occurrence of Precious Stones in the Drift." *Proceedings of the Iowa Academy of Science* 21: 203–204.

Nininger, H. H., and Addie Nininger. 1950. *The Nininger Collection of Meteorites*. American Museum, Winslow, Ariz.

Palache, Charles, Harry Berman, and Clifford Frondel. 1951. *The System of Mineralogy*. 7th ed., vol. 2. John Wiley and Sons, New York.

Pettijohn, Francis J. 1957. *Sedimentary Rocks*. Harper & Brothers, New York.

Sendlein, Lyle V. A. 1973. "Geology of the United States Gypsum Sperry, Iowa Mine." *Iowa Geological Survey Public Information Circular* 5: 67–87.

Sinotte, Stephen R. 1969. *The Fabulous Keokuk Geodes*, vol. 1, *Their Origin, Formation and Development in the Mississippian Lower Warsaw Beds of Southeastern Iowa and Adjacent States*. Wallace-Homestead Press, Des Moines, Iowa.

Spry, Paul G., and Keith B. Kutz. 1988. "A Fluid Inclusion and Stable Isotope Study of Minor Upper Mississippi Valley–Type Sulfide Mineralization in Iowa, Wisconsin and Illinois." In *New Perspectives on the Paleozoic History of the Mississippi Valley: An Examination of the Plum River Fault Zone*, ed. G. Ludvigson and B. Bunker. Great Lakes Section, Society of Economic Paleontologists and Mineralogists, 18th Annual Field Conference Guidebook, Iowa City.

Tester, Allen C. 1929. "The Dakota Stage of the Type Locality." *Iowa Geological Survey Annual Report* 35: 195–332.

Van Dorpe, Paul, and Mary A. Howes. 1986. "Mining Iowa's Coal Deposits." *Iowa Geology* 11: 12–16.

Van Tuyl, Francis M. 1922. "The Stratigraphy of the Mississippian Formations of Iowa." *Iowa Geological Survey Annual Report* 30: 304–346.

Wasson, J. T. 1985. *Meteorites*. W. H. Freeman, New York.

Witzke, B. J. 1987. "Geodes: A Look at Iowa's State Rock." *Iowa Geology* 12: 8–9.

Zeitner, June C. 1964. *Midwest Gem Trails*. 3d ed. Gembooks, Mentone, California.

2. Collecting Iowa's Minerals

Brannon, Joyce C., F. C. Podosek, and Roger K. McLimans. 1992. "Alleghenian Age of the Upper Mississippi Valley Zinc Lead Deposit Determined by Rb-Sr Dating of Sphalerite." *Nature* 356: 509–511.

Horick, Paul J. 1974. "The Minerals of Iowa." *Iowa Geological Survey Educational Series* 2.

———. 1988. "Rock and Mineral Collecting." *Iowa Geology* 13: 16–17.

Sinkankas, John. 1970. *Prospecting for Gemstones and Minerals.* Van Nostrand Reinhold, New York.

———. 1972. *Gemstone and Mineral Data Book.* Winchester Press, New York.

Weast, Robert, ed. 1984. *Handbook of Chemistry and Physics*, 56th ed. Chemical Rubber Company, Cleveland.

Zeitner, June C. 1964. *Midwest Gem Trails.* 3d ed. Gembooks, Mentone, California.

3. Occurrences of Iowa's Minerals

Anderson, Wayne I., and Paul L. Garvin. "The Cedar Valley Formation (Devonian), Black Hawk and Buchanan Counties: Carbonate Facies and Mineralization." *Geological Society of Iowa Guidebook* 42.

———, and Rick Stinchfield. 1989. "Pint's Quarry, Black Hawk County, Iowa." *Mineralogical Record* 20: 473–479.

Bard, Gary G. 1982. "Petrology and Diagenetic Features of the Fort Dodge Gypsum Beds." Ph.D. dissertation, Iowa State University, Ames.

Bleifuss, Rodney L. 1972. "The Iron Ores of Southeastern Minnesota." In *Geology of Minnesota*, ed. P. K. Sims and G. B. Morey. Minnesota Geological Survey, Centennial Volume, 498–505. St. Paul, Minnesota.

Brown, C. Ervin. 1967. "Fluorite in Crystal-lined Vugs in the Maquoketa Shale at Volga, Clayton County, Iowa." *American Mineralogist* 52: 1735–1750.

Calvin, Samuel. 1895. "Geology of Allamakee County." *Iowa Geological Survey Annual Report* 4: 35–120.

———, and H. Foster Bain. 1899. "Geology of Dubuque County." *Iowa Geological Survey Annual Report* 10: 379–652.

Cody, Robert D., Raymond R. Anderson, and Robert M. McKay. 1996. *Geology of the Fort Dodge Formation (Upper Jurassic), Webster County, Iowa.* Geological Survey Bureau, Iowa Department of Natural Resources Guidebook Series No. 19, Iowa City.

Dasenbrock, Frederick E. 1984. "Petrology and Diagenesis of the Fort Dodge Gypsum in the North Welles Quarry." Master's thesis, Iowa State University, Ames.

Dorheim, F., D. Koch, M. Parker, B. Peterson, J. Prior, R. Runvik, and L. Sendlein. 1972. "Tour of the U.S. Gypsum Company's Sperry Mine and Plant." *Eighth Forum on the Geology of Industrial Minerals, Field Trip Guidebook* 10–27.

Galpin, Sidney L. 1922. "The Rockford Geodes." *Proceedings of the Iowa Academy of Science* 29: 1922.

Garvin, Paul L. 1982. "Sulfide Mineralization at Mineral Creek Mines, Allamakee County, Iowa." *Proceedings of the Iowa Academy of Science* 89: 44–49.

———. 1984a. "Hydrothermal Mineralization of the Mississippi-Valley Type at the Martin-Marietta Quarry, Linn County, Iowa." *Proceedings of the Iowa Academy of Science* 91: 70–75.

———. 1984b. "Mineralization at Conklin Quarry." In *Underburden/Over-burden: An Examination of Paleozoic and Quaternary Strata at the Conklin Quarry Near Iowa City*, ed. B. Bunker and G. Hallberg. Geological Society of Iowa Guidebook 41.

———. 1995. "Paleokarst and Associated Mineralization at the Linwood Mine, Scott County, Iowa." *Journal of the Iowa Academy of Science* 102: 1–21.

———, and David M. Crawford. 1992. "The Minerals of the Linwood Mine, Scott County, Iowa." *Mineralogical Record* 23: 231–238.

———, and Gregory A. Ludvigson. 1988. "Mineralogy, Paragenesis and Stable Isotopic Compositions of Mineral Deposits Associated with Late Paleozoic Karst Fills in Johnson County, Iowa." In *New Perspectives on the Paleozoic History of the Mississippi Valley: An Examination of the Plum River Fault Zone*, ed. G. Ludvigson and B. Bunker. Great Lakes Section, Society of Economic Paleontologists and Mineralogists, 18th Annual Field Conference Guidebook, Iowa City.

———, and ———. 1993. "Epigenetic Sulfide Mineralization Associated with Pennsylvanian Paleokarst in Eastern Iowa." *Chemical Geology* 105: 271–290.

———, ———, and Edward M. Ripley. 1987. "Sulfur Isotope Reconnaissance of Minor Metal Sulfide Deposits Fringing the Upper Mississippi Valley Zinc-Lead District." *Economic Geology* 82: 1386–1394.

Gricius, A. 1964. "Preliminary Report on the Mineralogy of the Septaria Found Near Oskaloosa and Knoxville, Iowa." *Pick and Dopstick, Bulletin of the Chicago Rocks and Minerals Society* 19.

Hayes, David T. 1986. "Origin of Fracture Patterns and Insoluble Minerals in the Fort Dodge Gypsum, Webster County, Iowa." Master's thesis, Iowa State University, Ames.

Hayes, John B. 1964. "Geodes and Concretions from the Mississippian Warsaw Formation, Keokuk Region, Iowa, Illinois, Missouri." *Journal of Sedimentary Petrology* 34: 123–133.

Heyl, Allen V., Allen F. Agnew, Erwin J. Lyons, and Charles H. Behre Jr. 1959. "The Geology of the Upper Mississippi Valley Zinc-Lead District." *United States Geological Survey Professional Paper* 309.

Horick, Paul J. 1974. "The Minerals of Iowa." *Iowa Geological Survey Educational Series* 2.

Howell, Jesse V. 1915. "The Iron Ore Deposits Near Waukon, Iowa." *Iowa Geological Survey Annual Report* 25: 33–102.

Keyes, Charles R. 1893. "Annotated Catalogue of Minerals." *Iowa Geological Survey Annual Report* 1: 181–196.

Kuntze, Otto. 1899. "Occurrence of Quenstedtite Near Montpelier, Iowa." *American Geologist* 23: 119–121.

Leonard, A. G. 1894. "Satin Spar from Dubuque." *Proceedings of the Iowa Academy of Science* 1: 52–55.

————. 1896. "Lead and Zinc Deposits of Iowa." *Iowa Geological Survey Annual Report* 6: 11–66.

Lin, Feng-Chih. 1978. "Minerals in Vugs at Pint's Quarry, Raymond, Iowa." *Proceedings of the Iowa Academy of Science* 85: 25–30.

Ludvigson, Gregory A. 1976. "LANDSAT-1 Identified Linears in Northeast Iowa and Southwest Wisconsin and Their Relation to the Ordovician Stratigraphy, Structure and Sulfide Mineralization in the Area." Master's thesis, University of Iowa, Iowa City.

————, and James A. Dockal. 1984. "Lead and Zinc Mining in the Dubuque Area." *Iowa Geology* 9: 4–9.

McCormick, George R., and G. Bryan Bailey. 1973. "Brown Calcite from Iowa." *Mineralogical Record* 4: 188–190.

Menzel, Muriel, and Marilyn Pratt. 1963. "Southwestern Iowa's Rice Agates." *Iowan* 12: 19–20.

————, and ————. 1964. "Minerals of Pint's Quarry." *Earth Science* March–April: 63–67.

————, and ————. 1965. "New Prizes for Rockhounds." *Iowan* 14: 42–44, 52.

————, and ————. 1968. "Pint's Quarry Minerals." *Gems and Minerals* Jan.: 24–28.

————, and ————. 1969. "Iowa's Gemstones: Lake Superior Agates." *Iowan* 17: 48, 53.

Savage, Thomas E. 1902. "Geology of Henry County." *Iowa Geological Survey Annual Report* 12: 287–302.

Sendlein, Lyle V. A. 1973. "Geology of the United States Gypsum Sperry, Iowa Mine." *Iowa Geological Survey Public Information Circular* 5: 67–87.

Shipton, W. D. 1916. "The Occurrence of Barite in the Lead and Zinc District of Iowa, Illinois and Wisconsin." *Proceedings of the Iowa Academy of Science* 22: 237–245.

Sinkankas, John. 1970. *Prospecting for Gemstones and Minerals*. Van Nostrand Reinhold, New York.

Sinotte, Stephen R. 1969. *The Fabulous Keokuk Geodes*, vol. 1, *Their Origin, Formation and Development in the Mississippian Lower Warsaw Beds of Southeastern Iowa and Adjacent States*. Wallace-Homestead Press, Des Moines, Iowa.

Spencer, A. C. 1895. "Certain Minerals of Webster County, Iowa." *Proceedings of the Iowa Academy of Science* 2: 143–145.

Spry, Paul G., and Keith B. Kutz. 1988. "A Fluid Inclusion and Stable Isotope Study of Minor Upper Mississippi Valley–Type Sulfide Mineralization in Iowa, Wisconsin and Illinois." In *New Perspectives on the Paleozoic History of the Mississippi Valley: An Examination of the Plum River Fault Zone,* ed. G. Ludvigson and B. Bunker. Great Lakes Section, Society of Economic Paleontologists and Mineralogists, 18th Annual Field Conference Guidebook, Iowa City.

Tripp, R. B. 1959. "The Mineralogy of Warsaw Formation Geodes." *Proceedings of the Iowa Academy of Science* 66: 350–356.

Udden, Johan A. 1898. "Geology of Muscatine County." *Iowa Geological Survey Annual Report* 9: 247–388.

Van Tuyl, Francis M. 1912. "Geodes of Keokuk Beds, Origin." *Proceedings of the Iowa Academy of Science* 19: 169–172.

———. 1922. "The Stratigraphy of the Mississippian Formations of Iowa." *Iowa Geological Survey Annual Report* 30: 304–346.

Wilder, Frank A. 1902. "Geology of Webster County." *Iowa Geological Survey Annual Report* 12: 63–236.

Witzke, Brian J. 1987. "Geodes: A Look at Iowa's State Rock." *Iowa Geology* 12: 8–9.

4. Iowa's Mineral Industries

Anderson, Raymond R. 1979. "The Future of Iowa's Mineral Industry." *Iowa Geology* 4: 11–12.

———. 1983. "In Search of Iowa Oil." *Iowa Geology* 8: 10–11.

———. 1987. "Gold in Iowa." *Iowa Geological Survey Bureau, Educational Materials* EM-16.

———. 1990. "Iowa's Deepest Well." *Iowa Geology* 15: 26–27.

———. 1992. "Oil Exploration in Iowa." *Iowa Geology* 17: 19.

———, and Billy J. Bunker. 1982. "Oil and Uranium Potential in Iowa." *Iowa Geology* 7: 14–16.

Avcin, Matthew J., Jr. 1979. "History of Coal Mining in Iowa." *Iowa Geology* 4: 12–13.

Bettis, E. Arthur, III, and William Green. 1993. "Uses of Geological Materials by Prehistoric Cultures." *Iowa Geology* 18: 8–13.

Beyer, Samuel W. 1909. "Peat Deposits in Iowa." *Iowa Geological Survey Annual Report* 19: 689–730.

———, and Ira A. Williams. 1904. "Technology of Clays." *Iowa Geological Survey Annual Report* 14: 148–318.

Bischoff, Harold L. 1979. "Clayton's Silica Mine." *Palimpsest* 55: 84–95.

Calvin, Samuel. 1895. "Geology of Allamakee County." *Iowa Geological Survey Annual Report* 4: 35–120.

———, and H. Foster Bain. 1900. "Geology of Dubuque County." *Iowa Geological Survey Annual Report* 10: 379–652.

Centerville Iowegian. 1958. "There Still Is Gold on the F. C. Orr Farm Here." December 27.

Cody, Robert D., Raymond R. Anderson, and Robert M. McKay. 1996. *Geology of the Fort Dodge Formation (Upper Jurassic), Webster County, Iowa*. Geological Survey Bureau, Iowa Department of Natural Resources Guidebook Series No. 19, Iowa City.

Cole, Cyrenus. 1938. *I Am a Man: The Indian Blackhawk*. State Historical Society of Iowa, Iowa City.

Cuthbertson, Richard D. 1957. "The History of Lead Mining in the Dubuque Area." Master's thesis, Drake University, Des Moines, Iowa.

Davis, Merle. 1990. "Horror at Lost Creek: A 1902 Coal Mine Disaster." *Palimpsest* 71: 99–117.

Des Moines Register. 1963. "There's Gold in Iowa Rivers." October 21.

Drake, Lon, and G. Todd Ririe. 1975. *Strip Mine Reclamation in South-Central Iowa.* Thirty-ninth Annual Tri-state Field Trip Guidebook, Iowa City, Iowa.

Eckel, Edwin C., and H. Foster Bain. 1905. "Cement and Cement Materials of Iowa." *Iowa Geological Survey Annual Report* 15: 37–124.

Galpin, Sidney L. 1925. "The Geology of the More Refractory Clays and Shales of Iowa." *Iowa Geological Survey Annual Report* 31: 53–90.

Garvin, Paul L. 1982. "Sulfide Mineralization at Mineral Creek Mines, Allamakee County, Iowa." *Proceedings of the Iowa Academy of Science* 89: 44–49.

———, and Orville J. Van Eck. 1978. "Strippable Coal Reserves in Twelve Southern Iowa Counties." *Proceedings of the Iowa Academy of Science* 85: 1–5.

Gilmore, Jack L. 1978. "Oil and Gas Exploration in Iowa." *Iowa Geological Survey Newsletter* 3: 18–19.

Goodwin, Cardinal. 1919. "The American Occupation of Iowa." *Journal of Iowa History and Politics* 17: 83–102.

Gordon, C. H. 1895. "Geology of Van Buren County." *Iowa Geological Survey Annual Report* 4: 121–196.

Grant, Stanley C. 1979. "Iowa's Mineral Heritage." *Iowa Geological Survey Open File Report*, November.

Gwynne, Charles S. 1943. "Ceramic Shales and Clays of Iowa." *Iowa Geological Survey Annual Report* 38: 263–378.

Hall, James, and J. D. Whitney. 1858. *The Geology Survey of the State of Iowa*, vol. 1, part 1.

Heyl, Allen V., Allen F. Agnew, Erwin J. Lyons, and Charles H. Behre Jr. 1959. "The Geology of the Upper Mississippi Valley Zinc-Lead District." *United States Geological Survey Professional Paper* 309.

Hinds, Henry. 1909. "Coal Deposits in Iowa." *Iowa Geological Survey Annual Report* 19: 21–396.

Howell, Jesse V. 1915. "The Iron Ore Deposits Near Waukon, Iowa." *Iowa Geological Survey Annual Report* 25: 33–102.

———. 1921. "Petroleum and Natural Gas in Iowa." *Iowa Geological Survey Annual Report* 29: 1–48.

Howes, Mary R. 1992. "Iowa Coal: Future for a New State's Growth." *Iowa Geology* 17: 16–17.

———, and Matthew A. Culp. 1989. "Underground Coal Mines of the Des Moines Area." *Iowa Geology* 14: 14–15.

Iowan. 1954. "The Ottumwa Gold Rush." *Iowan* 2: 40.

Kemmis, Timothy J. and Deborah J. Quade. 1988. "Sand and Gravel Resources of Iowa." *Iowa Geology* 13: 18–21.

Keyes, Charles R. 1894a. "Coal Deposits of Iowa." *Iowa Geological Survey Annual Report* 2: 1–536.

———. 1894b. "Geology of Des Moines County." *Iowa Geological Survey Annual Report* 3: 409–492.

———. 1912a. "Aboriginal Use of Mineral Coal and Its Discovery in the West." *Annals of Iowa* 10: 431–434.

———. 1912b. "Spanish Mines: An Episode in Primitive American Lead Mining." *Annals of Iowa* 10: 539–546.

———. 1913. "Historical Sketch of Mining in Iowa." *Iowa Geological Survey Annual Report* 22: 89–122.

Koch, Donald L. 1984. "Iowa's Mineral Resources." In *Status of Natural Resources in Iowa, 1984*. Iowa Natural Heritage Foundation, Des Moines.

Landis, Edwin R., and Orville J. Van Eck. 1965. "Coal Resources of Iowa." *Iowa Geological Survey Technical Report* 4.

Langworthy, Lucius H. 1854–1855. "Dubuque, Its History, Mines, Indian Legends, Etc." Lecture 1 (1854), Lecture 2 (1855), reprinted in the *Journal of Iowa History and Politics* 8: 366–422.

Lees, James H. 1909. "History of Coal Mining in Iowa." *Iowa Geological Survey Annual Report* 19: 521–590.

Leonard, A. G. 1896. "Lead and Zinc Deposits of Iowa." *Iowa Geological Survey Annual Report* 6: 11–66.

Ludvigson, Gregory A., and James A. Dockal. 1984. "Lead and Zinc Mining in the Dubuque Area." *Iowa Geology* 9: 4–9.

McKay, Robert M. 1985. "Gypsum Resources of Iowa." *Iowa Geology* 10: 12–15.

———. 1989. "Iowa's Cement Industry." *Iowa Geology* 14: 24–25.

———. 1992. "Mineral Production in Iowa." *Iowa Geology* 17: 13–15.

———. 1994. "The History of Nonfuel Mineral Resource Development in Iowa." Geological Survey Bureau, Iowa Department of Natural Resources, unpublished report.

———, and Michael J. Bounk. 1987. "Underground Limestone Mining." *Iowa Geology* 12: 24–26.

Mintz, Leigh W. 1977. *Historical Geology*, 2d ed. Charles E. Merrill, Columbus, Ohio.

Moir, W. J., ed. 1911. *Past and Present of Hardin County, Iowa*. B. F. Bowen, Indianapolis, Indiana.

Morrow, Toby. 1994. "A Key to the Identification of Chipped-Stone Raw Materials Found on Archaeological Sites in Iowa." *Journal of the Iowa Archeological Society* 41: 108–129.

Owen, David D. 1844. *Mineral Lands of the United States*. First Session, 26th United States Congress, Executive Document 239: 26–39.

Petersen, William J. 1931. "Perrot's Mines." *Palimpsest* 12: 405–413.

———. 1957a. "Quarrying in Iowa." *Palimpsest* 38: 177–208.

———. 1957b. "Forecasters of Quarrying." *Palimpsest* 38: 205–208.

———. 1958. "Gold on the Prairie." *Palimpsest* 39: 556–558.

———. 1963. "The Ottumwa Coal Palace." *Palimpsest* 44: 572–578.

Rehder, Denny, and Cecil Cook. 1972. *Grass Between the Rails: The Waukon,*

Iowa, Branch of the Milwaukee Railroad. Waukon and Mississippi Press, Des Moines, Iowa.

Rodenborn, Leo V. 1972. *Gypsum—The History of the Gypsum Industry in Fort Dodge and Webster County, Iowa.* Messenger Printing Co., Ames, Iowa.

Savage, T. E. 1905. "Geology of Fayette County." *Iowa Geological Survey Annual Report* 15: 433–546.

Schwieder, Dorothy. 1982. "Italian Americans in Iowa's Coal Mining Industry." *Annals of Iowa* 46: 263–278.

———. 1983. *Black Diamonds.* Iowa State University Press, Ames.

———, and Richard Kraemer. 1973. *Iowa's Coal Mining Heritage.* Department of Mines and Minerals, Des Moines, Iowa.

Shiras, Oliver P. 1902. "The Mines of Spain." *Annals of Iowa* 5: 321–334.

Stolp, R. N., and Frederick P. Deluca. 1976. *Perspectives on Iowa Coal.* Energy and Mineral Resources Research Institute, Iowa State University, Ames.

Swisher, Jacob A. 1945. "Mining in Iowa." *Journal of Iowa History and Politics* 47: 305–356.

Thompson, Carol A. 1992. "Peat Production and Protection in Iowa." *Iowa Geology* 17: 20–21.

Union Publishing Co. 1883. *History of Hardin County, Iowa.* Union Publishing Co., Springfield, Illinois.

Van der Zee, Jacob. 1915. "Early History of Lead Mining in the Iowa Country." *Journal of Iowa History and Politics* 15: 3–52.

Van Dorpe, Paul, and Mary A. Howes. 1986. "Mining Iowa's Coal Deposits." *Iowa Geology* 11: 12–16.

Waterloo Sunday Courier. 1973. "Gold in Iowa." July 1.

White, Charles A. 1870. *Report of the Geological Survey of the State of Iowa* 1: 1–391.

Wilder, Frank A. 1902. "The Geology of Webster County." *Iowa Geological Survey Annual Report* 12: 63–235.

———. 1919. "Gypsum: Its Occurrence, Origin, Technology and Uses." *Iowa Geological Survey Annual Report* 28: 1–558.

Wood, L. W. 1934. "Road and Concrete Materials of Southern Iowa." *Iowa Geological Survey Annual Report* 36: 15–310.

Worthen, A. H. 1858. "Geology of Certain Counties." *Geology of Iowa* 1: 132–258.

5. Stories of Iowa's Minerals

Allmann, R., and J. D. H. Donnay. 1969. "About the Structure of Iowaite." *American Mineralogist* 54: 296–299.

Anderson, Raymond R. "Meteorites in Iowa's History." *Iowa Geology* 18: 20–21.

Beitz, Ruth S. 1962. "Swarthy King of the Palace Age." *Iowan* 10: 42–45, 51.

Booth, Harry. 1990. "'You Got to Go Ahead and Get Killed': Lost Creek Remembered." *Palimpsest* 71: 118–125.

Brown, R. J. 1993. "P. T. Barnum Never Did Say 'There's a Sucker Born Every Minute.'" http://www.server.com/ephemera/library/refbarnum.html.

The Cardiff Giant. http://www.cardiffgiant.com/hello.html.

Cedar Rapids Evening Gazette. 1902. "Oskaloosa Mine Horror." January 25.

Davis, Merle. 1990. "Horror at Lost Creek: A 1902 Coal Mine Disaster," *Palimpsest* 71: 99–117.

Des Moines Daily News. 1902. "Thirty Dead and Twenty Are Maimed." January 24.

Downing, Glenn R. 1973. "Seeing the World in Grains of Sand." *Idaho State Journal*, July 6, Pocatello.

Dunn, James T. 1960. "The Cardiff Giant." *Iowan* 8: 10–13.

Gallaher, Ruth A. 1921. "The Cardiff Giant." *Palimpsest* 2: 269–281.

Heling, D., and A. Schwarz. 1992. "Iowaite in Serpentinite Muds at Sites 778, 779, 780, and 784: A Possible Cause for the Low Chlorinity of Pore Waters." *Proceedings of the Ocean Drilling Program, Scientific Results* 125: 313–323.

Henderson, Aileen K. 1986. "The Black Stones of Estherville." *Iowan* 34: 53–54, 56–59.

Hinrichs, Gustavus D. 1905. *The Amana Meteorites*. Carl Gustavus Hinrichs, St. Louis.

Hoffman, Philip. 1945. "The Lost Creek Disaster." *Palimpsest* 26: 21–27.

Kohls, D. W., and J. L. Rodda. 1967. "Iowaite, a New Hydrous Magnesium Hydroxide-Ferric Oxychloride from the Precambrian of Iowa." *American Mineralogist* 52: 1261–1271.

Kreiner, Carl B. 1922. "The Ottumwa Coal Palace." *Palimpsest* 3: 336–342.

Leonard, Arthur G. 1906. "Geology of Clayton County." *Iowa Geological Survey Annual Report* 16: 213–317.

New London Journal. 1967. "Iowa Legislature Names the Geode Iowa's 'State Rock.'" March 2.

Northwest Book and Job Establishment. 1870. *The Cardiff Giant Humbug*. Northwest Book and Job Establishment, Fort Dodge, Iowa.

Oskaloosa Daily Evening Gazette. 1902. "A Horrible Fate." January 24.

Ottumwa Daily Courier. 1890. Several articles on the Ottumwa coal palace during the period from September 16 to October 11.

Petersen, William J. 1963. "The Ottumwa Coal Palace." *Palimpsest* 44: 572–578.

Proctor, Sharon. 1967. "New Honors for Old Geodes: About Iowa's 'State Stone.'" *Davenport Times-Democrat*, March 12.

Rischmueller, Marian C. 1945. "McGregor Sand Artist." *Palimpsest* 26: 129–147.

Schwieder, Dorothy, and Richard Kraemer. 1973. *Iowa's Coal Mining Heritage*. Department of Mines and Minerals, Des Moines, Iowa.

Seifert, K., and D. Brunotte. 1996. "Geochemistry of Serpentinized Mantle Peridotite from Site 897 in the Iberia Abyssal Plain." *Proceedings of the Ocean Drilling Program, Scientific Results* 149: 413–424.

Toney, Perry S. 1964. "The Lost Gold Cave." *Iowan* 12: 12–13.

Wasson, J. J. 1985. *Meteorites*. W. H. Freeman, New York.

White, Andrew D. 1905. *The Autobiography of Andrew Dickson White*, vol 2. Century Co., New York.

Wilson, Ben Hur. 1927. "The Amana Meteor." *Palimpsest* 8: 379–390.

———. 1928. "The Estherville Meteor." *Palimpsest* 9: 317–333.

———. 1929. "The Forest City Meteor." *Palimpsest* 10: 145–155.

———. 1937. "The Marion Meteor." *Palimpsest* 18: 33–47.

———. 1944. "The Mapleton Meteorite." *Palimpsest* 25: 129–140.

———. 1958. "Visitors from Outer Space." *Palimpsest* 39: 145–208.

Witzke, Brian J. 1987. "Geodes: A Look at Iowa's State Rock." *Iowa Geology* 12: 8–9.

Wythe, Stella. 1953. "Pictured Rock Sand Potichomanie." *Spinning Wheel* May: 9, 35.

Index

Bold type indicates illustrations.

sulfur, 87

sulfuric acid, 38, 47, 48, 62, 130, 178, 217

synthetic mineral, 1

tektite, 7

topographic map, 56, **57**

Upper Mississippi Valley Zinc-Lead District, 26, 77, 79

vadose environment, 31

vivianite, 38

Volga exposure, 30, 112

Waterloo South Mine, 30, 48, 113

Wilson, Ben Hur, 185

Wyoming Hill, 38, 44

zinc mining industry, 154; mining methods, **154**

Bur Oak Books/Natural History

Birds of an Iowa Dooryard
By Althea R. Sherman

A Country So Full of Game:
The Story of Wildlife in Iowa
By James J. Dinsmore

Fragile Giants: A Natural History
of the Loess Hills
By Cornelia F. Mutel

Gardening in Iowa and
Surrounding Areas
By Veronica Lorson Fowler

Iowa Birdlife
By Gladys Black

The Iowa Breeding Bird Atlas
By Laura Spess Jackson,
Carol A. Thompson, and
James J. Dinsmore

Iowa's Minerals: Their Occurrence,
Origins, Industries, and Lore
By Paul Garvin

Landforms of Iowa
By Jean C. Prior

Land of the Fragile Giants:
Landscapes, Environments, and
Peoples of the Loess Hills
Edited by Cornelia F. Mutel and
Mary Swander

Okoboji Wetlands: A Lesson
in Natural History
By Michael J. Lannoo

Parsnips in the Snow: Talks with
Midwestern Gardeners
By Jane Anne Staw and Mary Swander

Prairies, Forests, and Wetlands:
The Restoration of Natural Landscape
Communities in Iowa
By Janette R. Thompson

Restoring the Tallgrass Prairie:
An Illustrated Manual for Iowa
and the Upper Midwest
By Shirley Shirley

The Vascular Plants of Iowa:
An Annotated Checklist and
Natural History
By Lawrence J. Eilers and
Dean M. Roosa

Weathering Winter: A Gardener's
Daybook
By Carl H. Klaus